Palaeogeography and Palaeobiogeography
Biodiversity in Space and Time

The Systematics Association Special Volume Series

Series Editor

David J. Gower
Department of Zoology, The Natural History Museum
London SW7 5BD, UK

The Systematics Association promotes all aspects of systematic biology by organising conferences and workshops on key themes in systematics, running annual
lecture series, publishing books and a newsletter, and awarding grants in support
of systematics research. Membership of the Association is open globally to profes-
sionals and amateurs with an interest in any branch of biology, including palaeobiology. Members are entitled to attend conferences at discounted rates, to apply
for grants, and to receive the newsletter and mailed information; they also receive
a generous discount on the purchase of all volumes produced by the Association.

The first of the Systematics Association's publications, *The New Systematics* (1940),
was a classic work edited by its then-president, Sir Julian Huxley. Since then, more
than 70 volumes have been published, often in rapidly expanding areas of science
where a modern synthesis is required.

The Association encourages researchers to organise symposia that result in mul-
tiauthored volumes. In 1997 the Association organised the first of its international
Biennial Conferences. This and subsequent Biennial Conferences, which are
designed to provide for systematists of all kinds, included themed symposia that
resulted in further publications. The Association also publishes volumes that are
not specifically linked to meetings, and encourages new publications (including
textbooks) in a broad range of systematics topics.

More information about the Systematics Association and its publications can be
found at our website: http://www.systass.org.

Other Systematics Association publications are listed after the index for this
volume.

The Systematics Association Special Volume Series 77

Palaeogeography and Palaeobiogeography

Biodiversity in Space and Time

Edited by

Paul Upchurch

Alistair J. McGowan

Claire S.C. Slater

CRC Press
Taylor & Francis Group
Boca Raton London New York

CRC Press is an imprint of the
Taylor & Francis Group, an **informa** business

A CHAPMAN & HALL BOOK

The cover illustration, designed by Alistair J. McGowan as the logo for the meeting Palaeogeography and Palaeobiogeography in Space and Time (Cambridge, UK, 2005), incorporates three palaeogeographic maps produced by the Paleogeographic Atlas Project (PGAP), Department of the Geophysical Sciences, University of Chicago. We gratefully acknowledge the permission of Professor David Rowley to use the maps on the cover.

CRC Press
Taylor & Francis Group
6000 Broken Sound Parkway NW, Suite 300
Boca Raton, FL 33487-2742

First issued in paperback 2016

© 2011 by Systematics Association
CRC Press is an imprint of Taylor & Francis Group, an Informa business

No claim to original U.S. Government works

Version Date: 2011914

ISBN 13: 978-1-138-19891-3 (pbk)
ISBN 13: 978-1-4200-4551-2 (hbk)

Library of Congress Cataloging-in-Publication Data

Palaeogeography and palaeobiogeography : biodiversity in space and time / editors, Paul
　　Upchurch, Alistair McGowan, Claire Slater.
　　　　　p. cm. -- (Systematics Association publications ; 77)
　　Includes bibliographical references and index.
　　ISBN 978-1-4200-4551-2 (hardcover : alk. paper)
　　1. Paleogeography. 2. Paleobiology. I. Upchurch, Paul. II. McGowan, Alistair. III.
Slater, Claire. IV. Title.

QE501.4.P3P226 2012
560--dc23　　　　　　　　　　　　　　　　　　　　　　　　　　2011036517

Visit the Taylor & Francis Web site at
http://www.taylorandfrancis.com

and the CRC Press Web site at
http://www.crcpress.com

Contents

Preface

Biogeography is the scientific discipline that is concerned with the geographic distribution of organisms across the Earth. Biogeography is a highly multidisciplinary area of science that draws on the fields of ecology, evolutionary biology, geography, statistics, and Earth sciences. Computing science, GIS, and remote sensing are playing increasingly important roles in the processing and display of biogeographic information. One of the key scientific aims of biogeography is to understand the processes responsible for generating these patterns, and to use this understanding to predict past and future distributions of organisms.

Many techniques, from a wide range of scientific disciplines, are used in biogeography, but the key information is an understanding of the biology and evolutionary relationships amongst the organisms being studied, and the accurate spatial and ecological description of the geographic areas where they live.

All evolutionary questions have a historical dimension, whether it is examining data from 50 years ago to understand shifting species ranges or using information from the fossil record to investigate how the positions of the continents and oceans have changed over the course of 500 million years. The time dimension has traditionally been the domain of geologists and palaeobiologists. The links between this group of scientists, who have usually trained in the Earth sciences, and Recent biogeographers and phylogeographers who are often from life sciences backgrounds are underdeveloped. We need to understand biodiversity in space and time, and to do so, these two groups need to work together. Cecca et al. (Chapter 1, this volume) argue that the dichotomy between neobiogeography and palaeobiogeography is a false one, so perhaps the notion of understanding biodiversity in all four dimensions would be an acceptable compromise.

Possibly one of the most divisive splits in biogeography is not along a split of neobiogeography versus palaeobiogeography; instead it is between ecological biogeography and historical biogeography. Morrone and Crisci (1995) in a review of historical biogeography presented a flowchart (their Figure 1) that had a high-level split between historical and ecological biogeography, but no commentary on methods and conflicts within ecological biogeography. The focus of that article was on historical biogeography; however, there was no sense that ecological approaches, which yield a wealth of data that can constrain models, could contribute to historical biogeography. Equally, when one is working on the ecology of a particular taxon or clade it can come as a rude shock when the findings of an historical analysis that ignores taxonomic identities exhibit findings that run counter to years of careful fieldwork. If we are to use all our data and skills as productively as possible in the service of conservation biology, we should collaborate across all scales in time and space.

To help forge links between biologists and ecologists working on the Recent flora and fauna and their colleagues working in 'deep time', palaeogeographers and palaeobiogeographers, we decided to organise a meeting where both groups could come together to discuss the common problems they face, and how the different data and analytical techniques available to different areas of biogeography can complement each other to enhance our understanding of the spatial aspects of evolution.

The notion of having a meeting that focused broadly on the relationship between palaeogeography and palaeobiogeography was something the editors discussed while working together between 2003 and 2005. The Second International Biogeography Society meeting in 2005 gave us a chance to gauge support for such a meeting. This volume presents a series of chapters that were presented, or incubated, at Palaeogeography and Palaeobiogeography: Biodiversity in Space and Time held at the Centre for Mathematical Sciences, Cambridge 10th–13th of April 2006. We had 17 presented talks starting with three keynotes and then the other talks were grouped into Palaeozoic, Mesozoic, Cenozoic, and Recent sessions. Although this division by geological time might seem to have reinforced the split between neo- and palaeobiogeography, we had a range of interests represented in the different sessions that drew out the common methodological interests that cut across disciplinary boundaries. The balance of posters was also slightly in favour of palaeontological subjects, but many of these posters focused on broader topics about how we might use biogeographic approaches to gain new insights into extant groups. Not every talk from the meeting was converted into a chapter in this book. Some were published elsewhere, but at least one collaborative chapter (McGowan and Neige, Chapter 7, this volume) arose as a result of the meeting. A number of new collaborations arose from the meeting as well. The abstract volume from the meeting can be obtained as a PDF by emailing Alistair McGowan (Alistair.McGowan@glasgow.ac.uk).

SUMMARY OF CHAPTERS

The chapters in this volume cover a range of topics but reflect some of the major overall themes of the meeting:

- Which approach(es) are best suited to reconstructing biogeographic histories under a range of circumstances?
- How do we maximise the use of organismal and Earth sciences data to improve our understanding of events in Earth history?
- How well do analytical techniques devised for researching the biogeography of extant organisms perform in the fossil record?
- Can alternative biodiversity metrics, particularly those based on morphological measurements, enhance our understanding of biogeographic patterns and processes?

Chapter 1 (Cecca et al.) presents the case that any distinction between neo- and palaeobiogeography is an artificial one that should be avoided. Instead, we should focus on the most reliable and up-to-date means of reconstructing area histories, which they propose should be based on discoverable patterns. They propose a more cautious means of introducing the time dimension into biogeographic analyses.

Chapter 2 (Brooks and Folinsbee) tackles one of the grand philosophical themes of the natural sciences: Should we prefer our models to be simple or complex? As biology is a notoriously complex and messy topic, with plenty of exceptions to the rules, they argue that complex models are a better mapping of reality than the austere simplicity of physics. Biogeographic methods should be only as simple as the input data allow. Of all the contributions in the volume, this chapter is the strongest attempt to integrate the two strands of biogeography into a whole, which has been a major project of Brooks during his research career. The workings of the PACT algorithm are further discussed using an example drawn from the hominoids.

Chapter 3 (Smith) reviews the uncertainties involved in building palaeogeographic reconstructions from one of the authors of some of the seminal work on plate tectonics (Bullard et al., 1965). This broad review covers a diverse range of topics: the geometric basis for calculating, the quality of palaeomagnetic datasets, the added uncertainties of generating pre-Jurassic palaeogeographic reconstructions, and the additional lines of Earth science evidence for choosing amongst potential reconstructions. Smith's contribution reminds us to be vigilant about potential errors in the reconstruction of plate positions and how such errors can be propagated into other aspects of palaeogeography, such as topography and coastline positioning and models of past climates. An awareness of these uncertainties should lead to more judicious use of palaeogeographic mapping products, as well as an understanding that biogeographic research also has a vital role to play in testing Earth science data.

Chapter 4 (Hafner and Riddle) provides one such test. This detailed analysis of the biogeography of taxa in the Peninsular, Sonoran, and Chihuahuan deserts brings in a wide range of data from the Earth sciences, including geomorphology and climatic data. The most surprising aspect of their findings is strong evidence across a range of taxa for previously unsuspected complexity in the geological assembly of the Baja California and Gulf of California, which has led them to call for more detailed palaeontological and geological investigation of the area on the basis of biological signals.

Chapter 5 (Stigall) reflects the growing interest in analysis of regional diversity dynamics in the fossil record. For approximately 30 years there was a huge interest in the analysis of global diversity trends in the fossil record throughout the entire Phanerozoic. This was a tremendously fruitful research programme, but in recent years there has been an increased use of a range of techniques and methods borrowed from ecology and conservation biology. Stigall has been one of the pioneers of this approach, and this chapter serves as both an exploration of how to understand the interplay of origination, extinction, and immigration with tectonics and eustatic sea-level change in Devonian age basins in eastern North America. The use of the algorithms developed by conservation biologists to compare predicted and observed occupancy in the fossil record is another development of great interest.

Chapter 6 (Mutter) reminds us that all analytical work in evolutionary biology is underpinned by first-class systematics. His contribution redescribes a group of early Mesozoic actinoptrygian fishes. However, rather than stop there, Mutter carries out a cladistics analysis of the taxa, proposing a new series of phylogenetic relationships amongst the taxa and, in turn, uses the taxon phylogenies as the basis of a biogeographic analysis. This is one of the few analyses that has used Liebermann's modified BPA technique, and this novel study makes a series of predictions about marine biogeographic events in the aftermath of the Permian/Triassic event that are likely to be of importance in the broader understanding of the recovery of the marine biota from the greatest mass extinction of all time.

Chapter 7 (McGowan and Neige) is an exploration of the potential for morphological information (disparity) to contribute extra information to biogeographic analyses. Neige extended his previous work on extant cuttlefishes and the spatial distribution of richness and disparity by bringing in evidence from a molecular phylogeny. His new analysis supports his earlier conclusions about certain regions being dominated by radiations of single clades, leading to high richness but low disparity. McGowan provides an example of how palaeobiology could benefit from spatially subdividing disparity data to identify regions where the amount of morphological variety is much greater or smaller than would be predicted by a given genus-richness. Further work that separates taxa into cosmopolitan and endemic groups tests whether regions with endemic taxa have high or low levels of disparity.

LITERATURE CITED

Bullard, E., Everett, J.E., and Smith, A.G. (1965) The fit of the continents around the Atlantic. *Philosophical Transactions of the Royal Society of London,* A258: 41–51.

Morrone, J.J. and Crisci, J.V. (1995) Historical biogeography: Introduction to methods. *Annual Review of Ecology and Systematics*, 26: 373–401.

Acknowledgements

First and foremost, we would like to thank all the contributors for providing the chapters and for their associated time and effort. We would also like to thank the Systematics Association and Taylor & Francis for publishing this volume.

The conference was made possible through financial and in-kind help from the following sponsors, and we are grateful to each of them for their contributions. The now-defunct National Institute for Environmental e-Science (NIEeS) gave us tremendous help with the logistics of the meeting and greatly assisted us by organising the venue, catering, and documentation, and providing experience and support. A major part of the UK e-Science agenda has been to have units that act to transform fields by providing support for meetings and e-Science collaborations, and we benefited considerably from this programme. The Quantifying and Understanding the Earth System (QUEST) programme and the Systematics Association provided significant funding. We acknowledge the assistance of The Curry Fund of the Geologists' Association and the International Palaeontological Association for their financial contributions toward the meeting. Taylor & Francis, Blackwell Publishing, and Cambridge University Press provided financial support, discounts, and journals and books. We also thank the Sedgwick Museum of Earth Sciences, University of Cambridge for providing a wonderful venue for the Icebreaker reception and Gonville & Caius College. We are also grateful to the students from the University of Cambridge who helped with the Icebreaker organization and other elements of the meeting.

About the Editors

Paul Upchurch, PhD, is senior lecturer in palaeobiology at the Department of Earth Sciences, UCL. Having obtained a PhD in vertebrate palaeontology from Cambridge in 1994, he used a series of fellowships from Sidney Sussex College (Cambridge), The Leverhulme Trust, and the NERC to develop his interest in historical biogeography. In particular, he has focused on the role that fossils can play in providing information on the timing and nature of biogeographic events such as vicariance and dispersal. Paul joined UCL in 2003 and has subsequently carried out research on dinosaur evolution, particularly with regard to their phylogenetic relationships and diversity patterns. At present he is working on a different, but related, aspect of deep time spatial distributions: latitudinal biodiversity gradients amongst terrestrial animals during the Cretaceous Period.

Alistair J. McGowan, PhD, FGS is currently based in the School of Geographical and Earth Sciences, University of Glasgow, funded by a Royal Society of Edinburgh/Scottish Government Postdoctoral Fellowship co-funded by Marie Curie Actions. He commenced his academic career as a geologist upon graduating from the University of Glasgow in geology and applied geology in 1994. After three years of assorted jobs, including rebuilding a wooden boat in the most remote peninsula on mainland Scotland, mountain footpath construction, and streetsweeping, he returned to academia in 1997 to study for a MSc in palaeobiology at the University of Bristol, where his interests in quantitative palaeobiology were able to develop. This led to five years in the Department of the Geophysical Sciences at the University of Chicago, where he completed his PhD in 2003. After a short spell at the Smithsonian National Museum of Natural History, he returned to the UK to work with Paul Upchurch at UCL on a biogeographic simulation project as a research associate. This formative period cemented his interest exploring biodiversity in all four dimensions. He is also a fellow of the British Trust for Ornithology and voluntary work on bird surveys has been important in developing his ideas about sampling problems and biases in the fossil record, as well as straddling the divide between ecological and historical biogeography.

Claire S. C. Slater, PhD, completed her undergraduate studies in natural sciences, specialising in geological sciences, in 2002 at the University of Cambridge. She completed a MSc at the Natural History Museum and Imperial College in London. She then returned to Trinity College, University of Cambridge, to commence her PhD in 2003. It was during her PhD that she had the great pleasure of being involved with co-organizing (with Alistair J. McGowan and Paul Upchurch) the conference from which this book stemmed. She is currently pursuing a career in law but retains an enduring interest in palaeontology, biogeography, and the sciences.

Contributors

Daniel R. Brooks
Department of Ecology and
   Evolutionary Biology
University of Toronto
Toronto, Ontario, Canada

Fabrizio Cecca
UPMC Université de Paris
Paris, France

Malte C. Ebach
School of Biological, Earth, and
   Environmental Sciences
University of New South Wales
and
Australian Museum
Sydney, Australia

Kaila E. Folinsbee
Department of Anthropology
Iowa State University
Ames, Iowa

David J. Hafner
Division of Mammals
Museum of Southwestern Biology
University of New Mexico
Albuquerque, New Mexico

Alistair J. McGowan
School of Geographical and Earth
   Sciences
University of Glasgow
Glasgow, United Kingdom

Juan J. Morrone
Museo de Zoología 'Alfonso L.
   Herrera'
Departamento de Biología Evolutiva
Universidad Nacional Autónoma de
   México (UNAM)
Mexico D.F., Mexico

Raoul J. Mutter
Bernard Price Institute for
   Palaeontological Research
University of the Witwatersrand
Johannesburg, South Africa

Pascal Neige
Centre des Sciences de la Terre and
   Biogéosciences
Université de Bourgogne
Dijon, France

Brett R. Riddle
School of Life Sciences
University of Nevada–Las Vegas
Las Vegas, Nevada

Alan G. Smith
Department of Earth Sciences
Sedgwick Museum
Cambridge, United Kingdom

Alycia L. Stigall
Department of Geological Sciences
and
Ohio Center for Ecology and
   Evolutionary Studies
Ohio University
Athens, Ohio

1 Biogeographical Convergence and Time-Slicing
Concepts and Methods in Comparative Biogeography

Fabrizio Cecca, Juan J. Morrone,
and Malte C. Ebach

CONTENTS

1.1 INTRODUCTION

Comparative biogeography (Parenti and Ebach, 2009) deals with biotic areas and their historical relationships. Biogeographical distributions are a result of processes acting at different temporal and spatial levels. Some authors define biogeography as the study of distributions of living organisms, whereas others recognise palaeobiogeography—the study of distributions of fossil taxa—as a different discipline (Patterson, 1981). We, however, agree with Rosen (1992) that neobiogeography and palaeobiogeography should be classified as parts of a common biogeographical aim rather than as separate fields with different goals. Palaeobiogeography explicitly includes time, whereas neobiogeography, a palaeontological term, does not integrate temporal data (*sensu* Upchurch and Hunn, 2002) or focuses on recent ecological factors or observable processes (i.e., ecological biogeography). The distinction between

1

palaeo- and neobiogeography, however, is based solely on its practitioners (palaeontologists and nonpalaeontologists) rather than on any solid biogeographical theory or method.

The palaeontologists Upchurch and Hunn (2002) considered that the extent to which temporal data are incorporated into biogeography depends on the researcher's choice regarding sources of spatial data and analytical methodologies. 'Geologically-oriented' and 'biologically-oriented' biogeographers actually may have different perceptions of time that clearly influence their approach to the problem. Some authors (Grande, 1985; Hunn and Upchurch, 2001; Upchurch and Hunn, 2002; Upchurch et al., 2002) have suggested that the temporal ranges of organisms are important because distribution patterns seem to 'decay' through time as new ones are superimposed. Donoghue and Moore (2003) postulated that cladistic biogeographical methods are susceptible to the confounding effects of 'pseudo-congruence' and 'pseudo-incongruence', if they do not incorporate information on the absolute timing of the diversification of the lineages. By 'pseudo-congruence' they referred to cases where different areagrams showed the same area relationships, although the taxa diversified at different times, presumably due to different underlying causes. 'Pseudo-incongruence' means that the areagrams show conflict but the ages of the taxa indicate that they diversified in response to different events.

We discuss herein the problem that may arise when periods of contact between biotas, previously endemic to particular areas, result in the erasure of patterns of endemism, making it more difficult to uncover area relationships even with rigorous procedures. We also discuss the use of time-based methods to address this problem.

1.1.1 TERMINOLOGICAL DIGRESSION

A serious obstacle to communication in biogeography is the way in which different authors use specialised jargon according to different 'schools'. Before discussing the main issues, we would like to clarify some key terms, discussed in Ebach and Humphries (2002) and Ebach and Williams (2004), which are based on a rigorous distinction between patterns and processes.

Diachrony (also *diachronic theory*) is a concept or method that is based on an ad hoc biogeographic event that is postulated by an hypothetical explanatory mechanism (i.e., jump dispersal). The main diachronies in systematics and biogeography are ancestor-descendant lineages and the centre of origin concept, respectively. 'Diachronic' herein has a different sense from 'diachronous' in geology or palaeontology.

Synchrony (also *synchronic theory*) is a concept or method that is based on processes that form patterns, which are discoverable. The main synchrony in both systematics and biogeography is that of systematic relationships (i.e., cladograms).

Proximal is a relative concept and refers to relationships between homologous characters that are not based on explanatory mechanisms. In the case of biogeography, a proximal relationship would state that an area A is closer to an area B rather than to an area C.

Geographic paralogy was defined by Nelson and Ladiges (1996) by analogy with its use in molecular systematics, where it refers to a misleading comparison between duplicated genes that have independent histories. An ideal taxon-area cladogram or areagram consists of each taxon occurring in a single endemic area. When two or more taxa occur in the same area (redundancy), nodes are deemed paralogous. Nelson and Ladiges (2001, p. 390) stated that 'a node of an areagram is either geographically paralogous or it is not', and 'any interpretation of a paralogous node is apt to be artefactual'. Geographical data associated with nonparalogous nodes are the only data relevant to comparative biogeography, so geographic paralogy must be removed from the data to prevent artefactual interpretations, at least to a significant degree.

Ambiguity (also *ambiguous distribution*) is produced by widespread taxa (i.e., taxa occurring in more than one area) and by duplication of areas in areagrams through geographic paralogy. Ambiguity is 'incongruence' and must either be resolved or ignored, because the aim of cladistic biogeography is to discover geographical congruence.

Geographical congruence is the most important concept in cladistic biogeography. It corresponds to the finding of 'congruent patterns' (i.e., analytical patterns that are identical between unrelated taxa and which are generally interpreted as having a common cause). Vicariance is herein defined as *disjunct distribution* or *geographical isolation*, which is the result of some processes rather than an explicit mechanism or event (contra Hovenkamp, 1997). The uplift of the Panama Isthmus produced general patterns of biogeographical convergence in the north–south exchange of different groups of continental organisms, as well as vicariance (disjunction) of different groups of marine organisms (Waller, 1969; Rosen, 1984). Congruence can only be found once an initial pattern has been established. In comparative biogeography, area homology (i.e., geographical congruence) is usually considered to be the result of vicariance. Geographical paralogy is not congruent, not informative, and cannot be used to make logical discoveries, thus the relationships within a taxon-area cladogram are not affected when biogeographical ambiguities (paralogous nodes and widespread taxa) are removed (Nelson and Ladiges, 1996, 2001; Ebach and Williams, 2004; Ebach et al., 2005; Parenti and Ebach, 2009).

Biogeographical convergence is produced when different biotas are put into contact (Hallam, 1974). It possibly results in dispersal and reticulated, nonhierarchical, evolution of biotas. Convergent biotas are analogous to hybrids in systematics and therefore cannot be resolved with cladistic techniques that are 'tuned' to the discovery of branching hierarchical relationships.

Dispersal is a term that has been used with different and ambiguous meanings, because it does not specify any particular process. It appears to be an 'explanation of the pattern in terms of ideas about process' (Eldredge, 1981). The very broad use and the different meanings of this term, particularly the lack of distinctions between long- and short-term processes, highlights the confusion existing in most dispersalist explanations (Cecca, 2002). Dispersal has been commonly used to embrace different biogeographical processes acting on different temporal scales, namely the routine transport of propagules (short-term, or biological timescale), the chance crossing of barriers (short- to long-term), and the change of the distributional area of a taxon (short- to long-term, or evolutionary timescale). The distinction proposed by Rosen (1988) between processes of dispersion, which correspond to 'maintenance processes' (part of ecology), and evolutionary patterns of dispersal, has received scant attention from biogeographers. Dispersal can also be interpreted as *dispersion* (*sensu* Platnick, 1976), namely the movement of an organism or biota within its area of distribution. Dispersion does not form patterns until the distribution is disrupted in some way that leads to geographical isolation (vicariance). In other words if we were to view dispersal as a form of dispersion, geographical congruence can only emerge once that distribution has undergone subsequent vicariance. Hypothetical mechanisms have been postulated with the aim of explaining the interaction between dispersion and vicariance. One of these is *geodispersal* (Lieberman and Eldrege, 1996), which postulates a succession of dispersion and isolation events.

1.2   BIOGEOGRAPHICAL CONVERGENCE

The use of temporal data is avoided in cladistic biogeography because of the risk of incorporating ideas about unobserved processes into the interpretation of biogeographic patterns. This would imply unverifiable assumptions, with the risk of falling back on narrative scenarios. The need for considering time in biogeography, however, becomes clearer in cases of biogeographical convergence. The terms 'convergence' and 'divergence' have been proposed by Hallam (1974) to distinguish two extreme biogeographical patterns. Widespread taxa and redundancy identify biogeographical convergence, whereas vicariance is the most common interpretation of divergence patterns. Convergence can be the result of area coalescence (as a result of elimination of geographical barriers). Analyses of biogeographical convergence are unlikely to show congruence.

Upchurch et al. (2002) noted that historical analyses over an extensive stratigraphical range may fail to find the correct area relationships. This point is illustrated by the hypothetical succession of area separations followed by area coalescence described by Upchurch and Hunn (2002), where branching relationships only are evident when extensive dispersal has not yet overwhelmed the original vicariant pattern. In fact, if the biotas of areas A and C, formerly separated by vicariance following the pattern A(BC), merge and the third biota of area B remains isolated, the final relation will be uninformative because we will have two biotas and two areas: A1 (= A + C) and B. Area coalescence causes vicariance patterns to fade out through time (Grande, 1985). Biogeographical convergence leads to a sort of biogeographical 'overprinting', thus analytical techniques based on parsimony algorithms are probably inappropriate (Young, 1995). Of course, previously merged biotas may subsequently undergo vicariance, as already stressed by Rosen and Smith (1988), therefore the question remains of how patterns other than congruent vicariant ones can be treated without a 'multiple time plane approach' (Rosen and Smith, 1988) or 'time-slicing' (Upchurch et al., 2002). The latter term corresponds to the analysis of biotic distributional data according to a sequence of individual stratigraphical intervals (time slices).

Let us consider an example (Figure 1.1): originally (time t_0), we have a biota A. Then a vicariant event at time t_1 produces two biotas endemic to areas B and C. At time t_2 these biotas converge into a single biota in area D because of geographical coalescence, but a trace of the previous endemism may be still recognisable. Finally, at time t_3 two new endemic biotic components occur in areas E and F after a further vicariant event which subdivided the former biota inhabiting area D. How does one recognise the vicariant event at t_1 if the effects of biogeographical convergence that

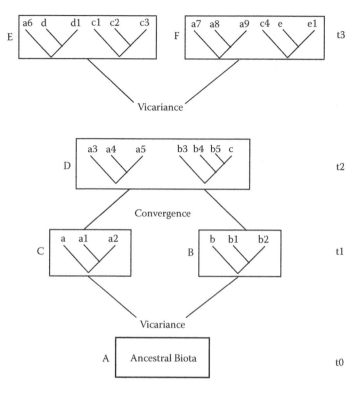

FIGURE 1.1 Hypothetical evolutionary history from time t_0 to t_3 implying a temporal succession of biogeographic divergence and convergence events. A–F: geographic areas; a–e: lineages.

formed the biota of area D overwhelmed the biotic fingerprint of the two original biotic components B and C? How does one recognise what happened at times t_1 and t_2 if one only studies the latest time plane (t_3) and the lineages occurring in areas E and F?

In order to minimise assumptions, a diachronic approach should be avoided. Some cladistic biogeographers address this problem in a synchronic (proximal) approach, eliminating paralogous distributions with subtree analysis and the transparent method (Ebach et al., 2005; *sensu* Nelson, 1984). Lieberman and Eldredge (1996) and Lieberman (2000) reintroduced the concept of biotic dispersal (Platnick and Nelson, 1978), which they called *geodispersal*. This term is synonymous and therefore redundant as the former takes precedence (see Parenti and Ebach, 2009). Unlike ad hoc dispersal, biotic dispersal may form congruent patterns, but it does not constitute an event, as biota are expanding their area of distribution. Furthermore, Lieberman and Eldredge (1996) used a modified version of Brooks Parsimony Analysis (BPA), a method that erroneously matches the nodes of a cladogram with the components of an areagram (Ebach et al., 2003, 2005).

Events of biogeographical convergence like those described for Mesozoic ammonites (Cecca, 2002) or for plant and animal taxa from the Mexican Transition Zone

(Halffter, 1987; Morrone, 2005a) produce a sort of 'overprinting' of past biogeographical histories by more recent patterns (e.g., reticulated area histories). This 'overprinting' reduces the chance of establishing area relationships through congruence, which is the final goal of comparative biogeography. We agree with Grande (1985) and Upchurch and Hunn (2002) that the solution to problems posed by instances of biogeographical convergence is time-slicing. We may use time-slicing when the fossil record is good enough (palaeobiogeography) or reliable molecular clocks are available (neobiogeography), and then apply a synchronic approach for each time slice (Morrone, 2009).

Although assessments of faunal similarity are usually undertaken with fauna of successive geological ages, traditional 'cladistic' biogeography has only used data on organism relationships and spatial distributions from a single time plane (usually the present). Time-slicing may reconcile the use of time and a synchronic approach. Ideally, palaeobiogeographers should be able to use a synchronic approach for each time slice they identify. This is difficult because of the limits imposed by geological constraints (e.g., insufficient precision or resolution of chronological correlations, incompleteness of the fossil record, etc.). It would be interesting, however, to unfold (through the analysis of distributional patterns) the possible sequence of the 'compressed' (i.e., temporally integrated) biotas within a single (most recent) time plane by interpreting the results of analyses of a succession of chronologically different datasets.

1.3 METHODS

1.3.1 Parsimony Analysis of Endemicity (PAE)

Originally proposed by Rosen (1984), it was first applied the following year (Rosen, 1985), and refined by Rosen (1988) and Rosen and Smith (1988) who published the first PAE areagrams. Originally devised for palaeobiogeography, PAE offers a possible analytical solution to the problem of time in biogeography. Rosen and Smith (1988) and Smith (1988) analysed time-sliced series of stratigraphically coeval data of fossil taxa to show changing relationships between sample localities and biotas over successive time horizons; however, PAE has commonly been used for a single time horizon (Cecca, 2002; Escalante and Morrone, 2003; Morrone, 2004, 2009). If a data matrix is established for each time slice and then processed to obtain area cladograms, a time-sliced series of PAE cladograms is obtained, where support for the nodes in one time slice can be looked for in cladograms of previous slices.

PAE *generates* patterns of endemicity based on similarity in cases where areagrams, and the means to derive them, are absent. PAE differs from cladistic biogeographical methods in that it uses sets of contemporaneous unrelated taxa (monophyletic groups are not essential although they are desirable for taxonomic precision and resolution) from one or more geological horizons (Rosen, 1988). PAE cladograms are hypotheses about similarities between areas of endemism based on biotic assemblages of sample areas, whereas area cladistic relationships are based on closely related biotas. A PAE cladogram is therefore an hypothesis about similarities between biotas and their respective areas of endemism. These similarities are interpreted as indicating the history of the most recent relative times in the past when particular pairs of sister sample areas were in biogeographical continuity. The historical succession of proposed biogeographical events can be compared with independent lines of evidence (e.g., geology, geophysics).

Several authors (Brooks and Van Veller, 2003; Escalante and Morrone, 2003; Morrone, 2004, 2005b, 2009; Porzecanski and Cracraft, 2005; Santos, 2005) have recently discussed PAE, which has been dismissed as a 'phenetic' technique by some phylogenetic biogeographers (Lieberman, 2000; Upchurch and Hunn, 2002; Brooks and Van Veller, 2003; Santos, 2005). PAE differs from Brooks Parsimony Analysis (Brooks et al., 2001) in that it uses the distributions of sampled taxa that may be monophyletic or not. Ebach et al. (2003, p. 1291) state that 'The theory and implementation [of both PAE and BPA], however, are exactly the same, with the exception that BPA and Assumption 0 attempt to insert a matrix that represents a fully resolved tree'. PAE does not introduce spurious data to resolve biogeographical ambiguities a posteriori; that is, it does not suffer from the problems outlined by Ebach and Humphries (2002) and Ebach et al. (2003) for BPA. According to Upchurch and Hunn (2002), PAE is a phenetic method because absences (being coded as '0') are assumed to be genuine instead of being treated as uncertainties (and hence coded as '?'). But would the coding of absences as uncertainties then make PAE a nonphenetic method? Upchurch (2004) also stated that areagrams produced by PAE may show erroneous area relationships when one of the areas analysed has undergone a more rapid extinction or evolution than the others. Extinction, however, deeply affects the whole suite of palaeobiogeographical information, regardless of the method used to analyse it. As yet, there are no techniques to overcome this problem.

Time-slicing has been under-explored by PAE, apart from Rosen and Smith's (1988) original paper and a further study by Smith (1988). From a pragmatic point of view, cladistic biogeography of invertebrate fossil groups can scarcely yet be started because the literature lacks cladistic analyses of fossil invertebrates (with the exception of trilobites: Lieberman, 2000; Ebach and Edgecombe, 2001). In addition, reticulated biotic evolution would be *ambiguous* for cladistic biogeography in that it would not produce congruence. At least theoretically, biogeographical convergence can be analysed through time-slicing, and PAE offers a way to analyse these ambiguous patterns.

1.3.2 Area Cladistics

Ebach (2003; see also Ebach and Edgecombe, 2001; Ebach and Humphries, 2002; Morrone, 2009) proposed the method called area cladistics, which is aimed at uncovering the proximal positions of the continents through time. It introduces time using a synchronic (proximal) approach, based on Young's (1987) *terrane analysis*, which selects a range of hypothetical proximal distance relationships between areas (geological cladograms) that best suit the resulting general areagram (geographical congruence). Consider the same space inhabited by two distinct biotas at two distinct times, for example, a present-day geographical area A containing both living and fossil taxa, and a stratigraphical section at a single locality A containing two fossil faunas of two distinct ages. With a synchronic approach, Ebach and Humphries (2002, Figure 6) proposed that these cases be analysed by considering A as two areas, A1 and A2, noting that 'Long-ranging taxa that inhabit two or more [habitats] in the same space can be treated as widespread taxa' (Ebach and Humphries, 2002, p. 435). In order to discover patterns of area relationship, time-slicing is needed for the area and not for the general areagram. This method avoids the generation of nonhierarchical or reticulate patterns. We are not aware of the analysis of cases of proven area coalescence and the resulting biogeographical convergence with area cladistics.

1.3.3   TEMPORALLY PARTITIONED COMPONENT ANALYSIS (TPCA)

Upchurch and Hunn (2002) proposed this method to incorporate explicitly temporal data into a cladistic biogeographical analysis (see also Hunn and Upchurch, 2001 and Upchurch et al., 2002). After *mass coherent dispersal*, resulting from area coalescence events (the same as biogeographical convergence), 'the histories of areas and biotas will have a reticulated rather than a branching structure, raising the question as to how well cladistic biogeographical techniques will be able to accurately analyse and depict a reticulate system' (Upchurch and Hunn, 2002, p. 280). In short, the starting point of TPCA is the existence of taxon-area cladograms for the taxonomic groups on which the analysis is based. The method comprises two analytical stages: (1) the finding of 'optimal area cladograms' (OACs) by determining which area relationships provide the 'best' (under optimality criteria) explanation for the spatial distributions observed in the taxon cladogram; and (2) a randomisation test to determine whether the degree of fit between area and taxon cladogram is greater than would be expected by chance. A sequence of time slices is defined from the fossil record, and for each horizon, taxa that did not exist at that horizon are deleted. Thus, the taxon cladogram is pruned or temporally partitioned (taxon deletions take place after the phylogenetic analysis, not before).

Although the introduction of time-slicing may tend to uncover reticulate histories, it is interesting to note that, ideally, synchronic relations would be found for each time slice on the basis of phylogenetic relations. TPCA is a pragmatic analytical method that allows reticulate histories to be explained, where assumptions are relatively minimised, and the area and lineage duplications that may be produced by BPA are avoided.

1.4   DISCUSSION

Comparative biogeography addresses biogeographical divergence, which is usually explained by vicariance, but is ineffective in cases of biogeographical convergence. Other analytical techniques, such as PAE and TPCA, which allow time-slicing to be performed, should be preferred. Whether PAE is a phenetic method or not, this does not weaken its advantages with respect to phenetic techniques based on similarity indices. This pragmatic view is liable to criticism: as correctly noted by Upchurch, '...the estimation of overall similarity between assemblages is such an inaccurate technique that it should never be used to assess the historical relationships between biotas and/or geographic areas' (Upchurch, 2004). We recognise that scarce data cannot justify pragmatism, being better to conclude that a certain problem is intractable. Nevertheless, there are problems that are still tractable even with 'inexact' tools, and in such cases we prefer to make some progress, instead of 'waiting for

Godot', and then test the results against independent lines of evidence or compare them with results obtained with other methods.

Donoghue and Moore (2003) discussed some ad hoc approaches that have been applied to integrate temporal information in specific analyses (Page, 1990; Huelsenbeck et al., 1997, 2000; Donoghue et al., 2001; Sanmartín et al., 2001), but we do not think that they represent explicit methods for time-slicing. Molecular clocks have not yet been used explicitly for time-slicing. A promising area for applying them would be to address the mixture of different biotic elements in the Mexican Transition Zone (Morrone, 2005a).

Biogeography provides hypotheses about the historical dynamics of biodiversity, which are intimately linked to the geographical dimension. This means that temporal data must be considered while trying to avoid a narrative approach and diachronic statements that imply unverifiable assumptions. The synchronic approach should be preferred as it minimises assumptions. Ideally, each time slice should be analysed and explained using a synchronic approach. The accurate analysis of historical successions of vicariant and convergent events through time-slicing is much needed for the study of biodiversity changes over time.

SUMMARY

The use of fossil data and molecular clocks in systematics highlights the necessity to introduce the concept of 'time-slicing' into comparative biogeography. Methods available for incorporating temporal information such as Parsimony Analysis of Endemicity (PAE), area cladistics, and Temporally Partitioned Component Analysis (TPCA), are discussed and compared to a synchronic approach. An analysis of these methods demonstrates the necessity for a synchronic approach and dispels the erroneous division between 'palaeo' and 'neo' biogeography.

ACKNOWLEDGEMENTS

We would like to thank Philippe Janvier (Muséum National d'Histoire Naturelle, Paris) and René Zaragueta (Université Pierre et Marie Curie, Paris) for interesting discussions and exchanges on the subject of this chapter. We are indebted to two anonymous referees for their constructive remarks that improved the manuscript.

LITERATURE CITED

Brooks, D.R. and Van Veller, M.G.P. (2003) Critique of parsimony analysis of endemicity as a method of historical biogeography. *Journal of Biogeography,* 30: 819–825.

Brooks, D.R., Van Veller, M.G.P., and McLennan, D.A. (2001) How to do BPA, really. *Journal of Biogeography,* 28: 345–358.

Cecca, F. (2002) *Palaeobiogeography of Marine Fossil Invertebrates: Concepts and Methods,* Taylor & Francis, London and New York.

Donoghue, M.J. and Moore, B.R. (2003) Toward an integrative historical biogeography. *Integrative and Comparative Biology,* 43: 261–270.

Donoghue, M.J., Bell, C.D., and Li, J. (2001) Phylogenetic patterns in northern hemisphere plant geography. *International Journal Plant Science,* 162: S41–S52.

Ebach, M.C. (2003) Area cladistics. *Biologist,* 50: 169–172.

Ebach, M.C. and Edgecombe, G.D. (2001) Cladistic biogeography: Component-based methods and paleontological application. In *Fossils, Phylogeny, and Form: An Analytical Approach* (eds. J.M. Adrain, G.D. Edgecombe, and B.S. Lieberman), Kluwer/Plenum Press, New York, pp. 235–289.

Ebach, M.C. and Humphries, C.J. (2002) Cladistic biogeography and the art of discovery. *Journal of Biogeography,* 29: 427–444.

Ebach, M.C. and Williams, D.M. (2004) Congruence and language. *Taxon,* 53: 113–118.

Ebach, M.C., Humphries, C.J., and Williams, D.M. (2003) Phylogenetic biogeography deconstructed. *Journal of Biogeography,* 30: 1285–1296.

Ebach, M.C., Humphries, C.J., Newman, R.A., Williams, D., and Walsh S.A. (2005) Assumption 2: Opaque to intuition? *Journal of Biogeography,* 32: 781–787.

Eldredge, N. (1981) Discussion of M.D.F. Udvardy's paper. In *Vicariance Biogeography: A Critique* (eds. G. Nelson and D.E. Rosen), Columbia University Press, New York, pp. 34–38.

Escalante, T. and Morrone, J.J. (2003) ¿Para qué sirve el análisis de parsimonia de endemismos? In *Una Perspectiva Latinoamericana de la Biogeografía* (eds. J.J. Morrone and J. Llorente Bousquets), Las Prensas de Ciencias, UNAM, Mexico, DF, pp. 167–172.

Grande, L. (1985) The use of paleontology in systematics and biogeography, and a time control refinement for historical biogeography. *Paleobiology,* 11: 234–243.

Halffter, G. (1987) Biogeography of the montane entomofauna of Mexico and Central America. *Annual Review of Entomology,* 32: 95–114.

Hallam, A. (1974) Changing patterns of provinciality and diversity of fossil animals in relation to plate tectonics. *Journal of Biogeography,* 1: 213–225.

Hovenkamp P. (1997) Vicariance events, not areas, should be used in biogeographical analysis. *Cladistics,* 13: 67–79.

Huelsenbeck, J.P., Rannala, B., and Larget, B. (2000) A Bayesian framework for the analysis of co-speciation. *Evolution,* 54: 352–364.

Huelsenbeck, J.P., Rannala, B., and Yang, Z. (1997) Statistical tests of host-parasite cospeciation. *Evolution,* 51: 410–419.

Hunn, C.A. and Upchurch, P. (2001) The importance of time/space in diagnosing the causality of phylogenetic events: Towards a "chronobiogeographical paradigm." *Systematic Biology,* 50: 391–407.

Lieberman, B.S. (2000) *Paleobiogeography: Using Fossils to Study Global Change, Plate Tectonics, and Evolution,* Kluwer Academic/Plenum, New York.

Lieberman, B.S. and Eldredge, N. (1996) Trilobite biogeography in the Middle Devonian: Geological processes and analytical methods. *Paleobiology,* 22: 66–79.

Morrone, J.J. (2004) *Homología Biogeográfica: Las Coordenadas Espaciales de la Vida.* Cuadernos del Instituto de Biología 37, Instituto de Biología, UNAM, Mexico, DF.

Morrone, J.J. (2005a) Hacia una síntesis biogeográfica de México. *Revista Mexicana de Biodiversidad,* 76: 207–252.

Morrone, J.J. (2005b) Cladistic biogeography: Identity and place. *Journal of Biogeography,* 32: 1281–1284.

Morrone, J.J. (2009) *Evolutionary Biogeography: An Integrative Approach with Case Studies,* Columbia University Press, New York.

Nelson, G. (1984) Cladistics and biogeography. In *Cladistics: Perspectives on the Reconstruction of Evolutionary History* (eds. T. Duncan and T.F. Stuessy), Columbia University Press, New York, pp. 273–293.

Nelson, G. and Ladiges, P.Y. (1996) Paralogy in cladistic biogeography and analysis of paralogy-free subtrees. *American Museum Novitates,* 3167: 1–58.

Nelson, G. and Ladiges, P.Y. (2001) Gondwana, vicariance biogeography, and the New York School revisited. *Australian Journal of Botany,* 49: 389–409.

Page, R.D.M. (1990) Temporal congruence and cladistic analysis of biogeography and cospe-
    ciation. *Systematic Zoology,* 39: 205–226.
Parenti, L.R. and Ebach, M.C. (2009) *Comparative Biogeography: Discovering and Classifying
    Biogeographical Patterns of a Dynamic Earth,* University of California Press, Berkeley.
Patterson, C. (1981) Methods of palaeobiogeography. In *Vicariance Biogeography: A Critique*
    (eds. G. Nelson and D.E. Rosen), Columbia University Press, New York, pp. 446–489.
Platnick, N.I. (1976) Concepts of dispersal in historical biogeography. *Systematic Zoology,*
    25: 294–295.
Platnick, N.I. and Nelson, G. (1978) A method of analysis for historical biogeography.
    Systematic Zoology, 27: 1–16.
Porzecanski, A.L. and Cracraft, J. (2005) Cladistic analysis of distributions and endemism
    (CADE): Using raw distributions of birds to unravel the biogeography of the South
    American aridlands. *Journal of Biogeography,* 32: 261–265.
Rosen, B.R. (1984) Reef coral biogeography and climate through the Late Cainozoic: Just
    islands in the sun or a critical pattern of islands? In *Fossils and Climate Geological
    Journal,* special issue II (ed. P. Brenchley), John Wiley and Sons, Chichester, UK, pp.
    201–262.
Rosen, B.R. (1985) Long-term geographical controls on regional diversity [lecture transcript
    summary]. *Journal of the Open University Geological Society,* 6: 25–30.
Rosen, B.R. (1988) Progress, problems, and patterns in the biogeography of reef corals and
    other tropical marine organisms. *Helgoländer Meeresuntersuchungen,* 42: 269–301.
Rosen, B.R. (1992) Empiricism and the biogeographical black box: Concepts and methods in
    marine palaeobiogeography. *Palaeogeography, Palaeoclimatology, Palaeoecology,* 92:
    171–205.
Rosen, B.R. and Smith, A.B. (1988) Tectonics from fossils? Analysis of reef-coral and sea-
    urchin distributions from late Cretaceous to Recent, using a new method. In *Gondwana
    and Tethys* (eds. M.G. Audley-Charles and A. Hallam), Geological Society Special
    Publication 37, London, pp. 275–306.
Sanmartín, I., Enghoff, H., and Ronquist, F. (2001) Patterns of animal dispersal, vicariance,
    and diversification in the Holarctic. *Biological Journal of the Linnean Society,* 73:
    345–390.
Santos, C.M.D. (2005) Parsimony analysis of endemicity: Time for an epitaph? *Journal of
    Biogeography,* 32: 1284–1286.
Smith, A.B. (1988) Late Palaeozoic biogeography of East Asia and palaeontological con-
    straints on plate tectonic reconstructions. *Philosophical Transactions of the Royal
    Society of London,* A236: 189–227.
Upchurch, P. (2004) Plunging in bravely [book review]. *Systematics and Biodiversity,* 1:
    515–517.
Upchurch, P. and Hunn, C.A. (2002) "Time": The neglected dimension in cladistic biogeogra-
    phy? In *Geobios,* 35 (*mémoire spéciale* 24) (eds. P. Monegatti, F. Cecca, and S. Raffi),
    pp. 277–286.
Upchurch, P., Hunn, C.A., and Norman, D.B. (2002) An analysis of dinosaurian biogeogra-
    phy: Evidence for the existence of vicariance and dispersal patterns caused by geologi-
    cal events. *Proceedings of the Royal Society of London,* B269: 613–621.
Waller, T.R. (1969) The evolution of the *Argopecten gibbus* stock (Mollusca: Bivalvia), with
    emphasis on the Tertiary and Quaternary species of Eastern North America. *Journal of
    Paleontology,* 43, special memoir 3: 1–125.
Young, G.C. (1987) Devonian palaeontological data and the Armorica problem.
    Palaeogeography, Palaeoclimatology, Palaeoecology, 60: 283–304.
Young, G.C. (1995) Application of cladistics to terrane history—Parsimony analysis of quali-
    tative geological data. *Journal of Southeast Asian Earth Sciences,* 11: 167–176.

2 Phylogenetic Methods in Palaeobiogeography
Changing from Simplicity to Complexity without Losing Parsimony

Daniel R. Brooks and Kaila E. Folinsbee

CONTENTS

2.1 INTRODUCTION

The development of phylogenetic analysis (Hennig, 1966; Wiley, 1981) and widespread adoption of quantitative methods to achieve those goals (see Brooks et al., 2007 for a review) catalysed a revolution in systematic biology. It also produced a travelling wave of advances and controversies permeating all of comparative evolutionary biology (Brooks and McLennan, 1991, 2002; Harvey and Pagel, 1991). Historical biogeography has not escaped this phenomenon. Since the early 1970s, historical biogeographers have found ample material for loud arguments over conceptual issues, methods, and results. The arguments almost always involve conflicts between simplicity and complexity, and have produced much heat but little light, and nothing approaching a synthesis. Rather, there is a 'simplicity group' (cladistic biogeography) and a 'complexity group' (phylogenetic biogeography; see Dowling et al., 2003; Ebach and Morrone, 2005; Van Veller et al., 2002). Both sides argue past each other, unable to comprehend the other side's perspective or anxieties, because they come from different ontological foundations (Van Veller et al., 2002). The emotional content of this conflict indicates that the basis for it is more fundamental than comparative biology. At the centre, it is a divide over whether one believes that dispersal events can ever produce congruent patterns across clades. The importance of dispersal versus vicariance is central to this divide; those who believe patterns can be best explained by invoking only vicariance (e.g., the simplicity perspective) seek to minimise the occurrences of dispersal, and those who believe dispersal may contribute to general patterns (e.g., the complexity perspective) seek to differentiate the two.

Anyone who has observed wild primates understands that we are descended from timorous beasts, and with good reason. Our evolutionary legacy did not include exceptional physical weapons, or great reproductive capacity, so we could not overpower or out-reproduce our surroundings. Our survival was based on out-thinking the competition. This placed a premium on accurately perceiving and generalising the complexity of our surroundings. Underestimate the complexity of your surroundings, and you are lunch; overestimate them, and fear makes you so indecisive that you starve. Working on an ancestral legacy of cleverness, natural selection seems to have succeeded remarkably. Humans have an excellent ability to perceive complex patterns. In fact, this is one type of task at which humans are still better than computers; efforts such as object-oriented programming are driven by the desire to help computers emulate human pattern recognition. And yet, we tend to distrust complex explanations, even for complex data. Why? Successful generalising gave the earliest humans a sense of security, so they learned to trust generalising. Good became associated with generalisation, evil with contingency, the complexity that cannot be generalised, that which does us harm, which frightens us. We distrust complexity because that is where the bogeymen are.

Initially, humans thanked supernatural forces for the generalities. They thanked supernatural forces for giving them the ability to understand their surroundings. Then they began to use these generalisations to manipulate and control the surroundings themselves, and thanked the supernatural forces for giving them the ability to control their surroundings. It was not until the Enlightenment that humans, in the western world at least, discarded the supernatural and decided to thank their own cleverness for the ability to control their surroundings. And yet we did not let complexity escape our history; good became associated with 'laws', and 'evil' became associated with 'contingency'. The superstitious fear of complexity, an historical constraint on our thinking, created a schism between what is and what ought to be, or between experience and reason (see Neiman, 2002 for an extended discussion). The Cartesian perspective was one in which the world ought to be reasonable (the deity that created the world must be reasonable), therefore it is reasonable. The appearance of unreasonableness is due to the activities of evil in the world, clouding our reason. The successful philosopher/scientist is one who is able to use human reason to see the underlying supernatural reason. The Newtonian perspective extended the Cartesian view, suggesting that the world ought to be not just reasonable, but also efficient; therefore, the universe is simple, a clockwork mechanism ruled by a small number of universal laws. The appearance of complexity and contingency is due to incomplete knowledge. Many biologists today associate Voltaire's *Candide* with the famous article on Panglossian adaptationism by Gould and Lewontin (1979), but *Candide* was a caricature of the Newtonian Revolution, not of neo-Darwinism.

The poster boy for the Newtonian Revolution was the principle of parsimony as articulated by Aristotle, that 'nature operates in the shortest way possible' and 'the more limited, if adequate, is always preferable'. The principle is also linked with the English philosopher and Franciscan monk William of Ockham, who advocated what is known as 'Ockham's Razor': *'pluralitas non est ponenda sine neccesitate'* ('plurality should not be posited without necessity') and *'non sunt multiplicanda entia praeter necessitatem'* ('entities should not be multiplied unnecessarily'). In this sense, the principle of simplicity obliges us to favour theories or hypotheses that make the fewest unwarranted, or ad hoc, assumptions about the data from which they are derived. This epistemological use of the principle does not necessarily imply that nature itself is parsimonious. Nonetheless, most scientists even today follow Newton and conduct their research as if they believe that nature is parsimonious in some sense, and they place much more credence in simple than in complex theories.

Despite the best efforts of philosophers for more than 700 years, however, no necessary link between parsimony and truth has ever been established. This would not be a problem if the empirical sense data of the world did not regularly conflict with what our reason tells us the world ought to be. The German philosopher Immanuel Kant spent his career worrying about this regular disagreement between experience and reason, between what we think *is* and what we think *ought to be*. When reason and experience conflict, which do we choose? Most western philosophers and scientists convinced themselves that the truth is simple and that the simple truth is always somewhere in the data. They also became accustomed to the idea that it sometimes takes complex manipulations of the data to find the simple truth. Kant, however, believed there was no single answer to this conundrum. He suggested that the conundrum itself is the source of tension between theory and data, which stimulates scientists to try to narrow the gap between 'is' and 'ought'.

The same conundrum faces us today in many areas of evolutionary biology, where we use the term 'model' as a synonym for 'reason'. When data and models conflict, what do we believe, and what do we do about it? Reason the data into agreement with the model or modify the model to fit the data? We all know that appearances can be deceiving, but sometimes our reason can falter as well. As Kant suggested, we need to narrow the gap between data and models.

Kluge (1997a, 1997b) suggested that such a process is deeply embedded in the philosophy and methodology of phylogenetic systematics. His notion of alternating cycles of discovery and evaluation, which Hennig (1966) called 'reciprocal illumination', is a means of basing our explanations, both corroboration of models and refutation of models, on the mutual agreement of data taken from as many different sources as possible. The nineteenth-century philosopher William Whewell called this the 'consilience of inductions' in a book published the same year Darwin and Wallace presented their joint findings on evolution to the Royal Society (Whewell, 1858).

2.2   BIOLOGY AS COMPLEXITY SCIENCE

The cosmologist Stephen Hawking has dubbed the twenty-first century the century of complexity, leading a parade of physicists who began to think about ontological complexity in the latter decade of the twentieth century. Biology ought to be well poised to take advantage of this sea change in our understanding of the nature of science and the universe, having been based on the ontology of complexity called Darwinism for almost 150 years, but it is not. Few fully realise that Darwinism is not a simple theory:

> ... [T]here are two factors: namely, the nature of the organism and the nature of the conditions. The former seems to be much more the important; for nearly similar variations sometimes arise under, as far as we can judge, dissimilar conditions; and, on the other hand, dissimilar variations arise under conditions which appear to be nearly uniform. (Darwin, 1872)

Most evolutionary biologists (and historians of biology) consider this passage no more than a general repudiation of Lamarckism. An alternative view, however, is that it is one of the first articulations of a complex theory: Darwin proposed that evolution is an emergent property of asymmetrical interactions between two different causal agents, each with its own properties, producing outcomes that are not readily predictable from knowledge of the properties of either agent, and thus historically contingent.

Darwin thought that organisms were historically and developmentally cohesive wholes, and therefore it was in the 'nature of the organism' to produce offspring that were all highly similar (but not identical) to each other and to their parents and other ancestors. He also postulated that reproduction produced variation without regard for environmental conditions, and therefore it was in the 'nature of the organism' to produce these offspring in numbers far exceeding the resources available for their support. When this inherent overproduction produced variety in critical characters, natural selection would preserve the versions that were functionally superior in that particular environmental context (adaptations). Whenever an environment changed, those organisms that already had the adaptations necessary to survive would do so, whereas those lacking appropriate adaptations would not. The production of organismal diversity thus required that organisms be at once autonomous from, and sensitive to, the environment, another example of complexity. Finally, Darwin visualised his viewpoint about the complexity of evolution with two metaphors, the phylogenetic tree and the tangled bank. The phylogenetic tree points to complexity arising from irreversible phenomena. By referring to species as 'communities of descent', and placing them in a single 'Tree of Life', Darwin emphasised that the fundamental explanatory principle is shared history. The tangled bank, by contrast, points to complexity arising from biological associations that do not share the same temporal and spatial origins, producing reticulated (tangled) patterns of descent.

This non-Newtonian perspective led many to accept Darwin's general proposition that evolution had occurred, but also to expend considerable effort trying to 'fix' Darwinism, making it more Newtonian (Depew and Weber, 1995). The neo-

Darwinian revolution was highly successful in producing a simple perspective on evolution that retained the essence of Darwinism, but only at the cost of blurring the distinction between Darwinian and Lamarckian perspectives on adaptation (see Brooks and McLennan, 2002 for a review and references). So powerful is this reliance on simplicity that proposals by biologists calling attention to the similarities between the complexity of Darwinism and the emerging acceptance of complexity on the part of physicists in the past 20 years have generally been met with hostility or embarrassed silence. Perhaps now that the physics community has begun to embrace complexity, more biologists will rediscover the Darwinian panorama, and overcome their fear of complexity as advocated so eloquently by the palaeontologist G. G. Simpson more than 50 years ago:

> ... [I]n every part of the whole, wonderful history of life, all the modes and all the factors of evolution are inextricably interwoven. The total process cannot be made simple, but it can be analyzed in part. It is not understood in all its appalling intricacy, but some understanding is in our grasp, and we may trust our own powers to obtain more. (Simpson, 1953)

2.3 HISTORICAL BIOGEOGRAPHY AS COMPLEXITY SCIENCE

> ... In considering the geographic distribution of organic beings over the face of the globe ... [w]e see ... some deep organic bond, through space and time, over the same areas of land and water, independently of physical conditions. The naturalist must be dull who is not led to enquire what this bond is. (Darwin, 1872)

Tangled bank complexity produces reticulated historical relationships in addition to relatively simple linear ones. We now know that methods of historical biogeographical analysis based on phylogenetic tree-based reasoning produce internally inconsistent results in direct proportion to the degree of tangled bank complexity in the data, an indication that the methods have oversimplified the data (Dowling, 2002; Dowling et al., 2003; Van Veller and Brooks, 2001; Van Veller et al., 1999, 2000, 2002, 2003). Reticulated area relationships are the rule, not the exception, because life has overrun the planet multiple times. That is, the planet is not made up of a grid, with each cell containing one species or one clade. Dispersal as well as isolation events produce general patterns in both space and time. Whenever methods that can recover both isolation and dispersal patterns have been applied, both classes of biogeographic processes have been shown to operate (e.g., Hembree, 2006; Lieberman and Eldredge, 1996; Rode and Lieberman, 2005a, 2005b).

Methods of analysing data, including those based on reasonable models (e.g., Ronquist, 1997), that do not allow for the possibility of reticulated area relationships and general patterns of dispersal, are oversimplifications and therefore obsolete anachronisms. The path toward developing a new generation of analytical methods is bounded by two complementary considerations: (1) we have overgeneralised when the cost of producing a simple explanation from the data outweighs the cost of a complex explanation; and (2) we have undergeneralised when we think there ought to be generalisations and we do not find any. Therefore, we need methods that generalise as much as the data allow (epistemological parsimony), not as much as we can imagine (ontological parsimony).

2.3.1   Phylogenetic Analysis for Comparing Trees, or PACT

PACT (Wojcicki and Brooks, 2004, 2005) is based on the assumption that tangled bank complexity is based on individual parts that each conform to relatively simple treelike complexity which, when combined, may nonetheless produce complex patterns of both general and unique relationships, including the possibility of reticulated relationships. PACT uses three guiding principles: (1) Assumption 0 (Wiley, 1986, 1988a, 1988b; Zandee and Roos, 1987), which states that in order not to oversimplify, you must analyse all data in each input string without modification, and your final analysis must be logically consistent with all input data; (2) the duplication rule (Brooks and McLennan, 2002): Assumption 0 is violated by analysis by methods of simplicity when the data result from reticulated processes. Oversimplification can be avoided (i.e., Assumption 0 can be satisfied) in such cases by duplicating entities with reticulated histories; and (3) epistemological parsimony (Brooks and McLennan, 2002): in order to find the maximum generalities supported by the data, entities should not be duplicated beyond necessity. Therefore, we should make only enough duplications to satisfy Assumption 0. Underscoring its connection with the ontology of complexity, note that Assumption 0 is more fundamental than parsimony. If evolutionary patterns were simple, Assumption 0 would never be violated, so we would never need to duplicate entities, and would never need parsimony. What we are seeking is the most general explanation possible given the data (i.e., the ontology of complexity: Van Veller and Brooks, 2001; Brooks and McLennan, 2002), not the simplest explanation we can imagine (i.e., the ontology of simplicity: Van Veller and Brooks, 2001; Brooks and McLennan, 2002). What we have to guard against is the use of models and analytical methods that oversimplify the data in an effort to generalise. That is, simplicity is not always the most parsimonious depiction of the real world (Van Veller and Brooks, 2001).

PACT utilises any data that can be represented as hierarchical strings, ranging from gene trees and other character state trees to species phylogenies. For historical biogeography, a minimum of three phylogenies is required (Brooks and McLennan, 2002). Wojcicki and Brooks (2004, 2005) recognised that nested (historical) hierarchical patterns in nature are combinations of only six different classes of pattern modules. Analysis of a set of data is accomplished by selecting any one of the input data strings as the initial template. Next, the final analysis is built by sequentially aligning each additional data string with the template. Thereafter, combinations are made to satisfy Assumption 0 using four combination rules. The combination rules are illustrated in Figures 2.1 to 2.6 and explained briefly in the figure legends; for more details, see Wojcicki and Brooks (2004, 2005), Brooks and Folinsbee (2005), Folinsbee and Brooks (2007), and Brooks and Van Veller (2008).

Elucidating complex evolutionary patterns using PACT comes with a cost of sorts. Explaining the complexity of a set of observations can only be achieved by abandoning a priori expectations of mechanism. For example, is the presence of a species in area F in Figure 2.3 an indication of a unique event involving the production of a

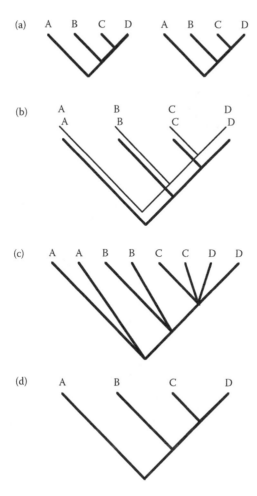

FIGURE 2.1 PACT for simple congruence. (a) Two taxon-area cladograms pertaining to areas A, B, C, and D. (b) The taxon-area cladograms aligned by common elements. (c) The taxon-area cladograms superimposed on each other. (d) PACT solution based on applying combination Rule 1 (Y + Y = Y).

novelty in one of the data strings, or is it an indication of a unique event involving loss from the other? PACT produces the same result regardless of the interpretation, so no assertions or assumptions about processes are embedded in the algorithm. For example, general patterns involving dispersal may result from movement of areas (geodispersal) or from movement of organisms (peripheral isolates speciation, recently rebranded as 'jump dispersal', e.g., Sanmartin, 2003; Sanmartin et al., 2007). The responsibility for making correct inferences from the patterns lies with the investigator, not the method for discovering and documenting patterns in the data. Similar reasoning can be applied to concepts such as lineage duplication, lineage sorting, and ancestral areas/hosts. All these concepts pertain to hypotheses of mechanism, and in order to maximise empirical robustness and avoid the possibility

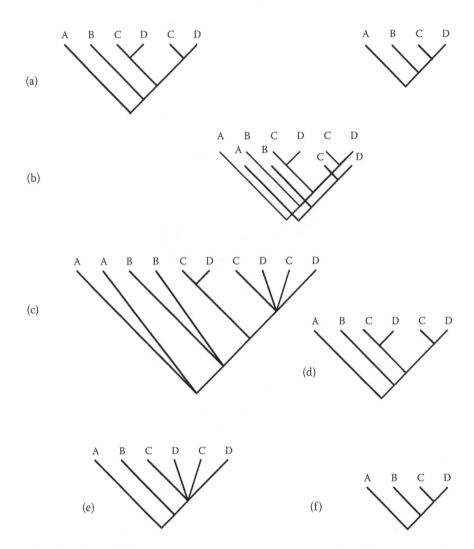

FIGURE 2.2 PACT for lineage duplication (sympatric speciation). (a) Two taxon-area cladograms pertaining to areas A, B, C, and D. (b) The taxon-area cladograms aligned by common elements. (c) The taxon-area cladograms superimposed on each other. (d) PACT solution based on applying combination Rule 1 (Y + Y = Y). Note that this leaves duplicate areas (CD) coming from a common node. (e) Two (CD) groups coming from a common node superimposed on each other. (f) PACT solution based on applying combination Rule 1 (Y + Y = Y).

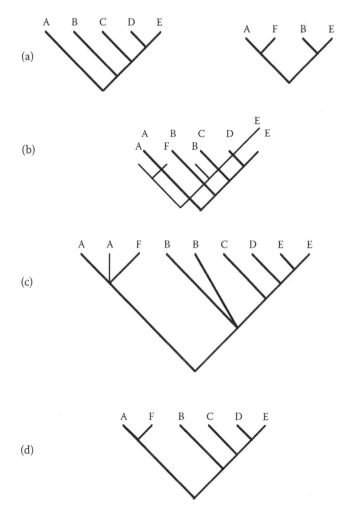

FIGURE 2.3 (a) Two taxon-area cladograms pertaining to areas A, B, C, D, E, and F. (b) The taxon-area cladograms aligned by common elements. (c) The taxon-area cladograms superimposed on each other. (d) PACT solution based on applying combination Rule 2 (Y + YN = YN). Y(A) + YN (AF) = YN (AF); Y (E) + YN (ED) = YN (ED); Y (B) + YN (C(DE)) = YN (B(C(DE))).

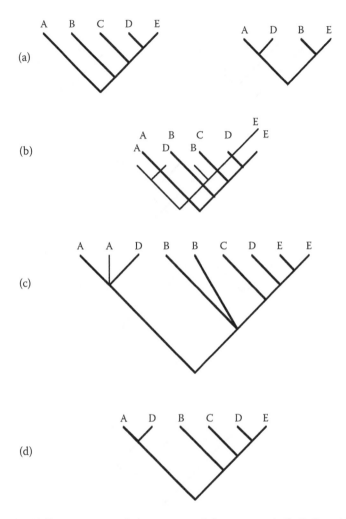

FIGURE 2.4   (a) Two taxon-area cladograms pertaining to areas A, B, C, D, and E. (b) The taxon-area cladograms aligned by common elements. (c) The taxon-area cladograms super-imposed on each other. (d) PACT solution based on applying combination Rule 2 (Y + YN = YN). Y (A) + YN (AD) = YN (AD); Y (E) + YN (ED) = YN (ED); Y (B) + N (C(DE)) = YN (B(C(DE))). Note that this example is the same as the one shown in Figure 2.3 with the exception that in this case we find evidence of reticulated relationships amongst areas.

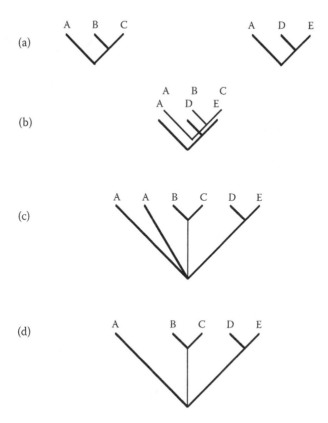

FIGURE 2.5   (a) Two taxon-area cladograms pertaining to areas A, B, C, D, and E. (b) The taxon-area cladograms aligned by common elements. (c) The taxon-area cladograms superimposed on each other. (d) PACT solution based on applying combination Rule 3 (YN + YN = YNN). Y(A)N(BC) + Y(A) N(DE) = YNN(A)(BC)(DE).

of circularity, all processes should be inferred following an analysis, not built into it. In much the same way, ACCTRAN and DELTRAN optimisation in phylogenetic methodology reveal different interpretations of the same data on a phylogenetic tree, but imply different evolutionary scenarios.

It is for this reason that PACT makes combinations beginning with the terminal branches. This avoids the assumptions, unintentionally embedded in all previous methods for historical biogeography, that all data strings associated today originated together at the same place and time, and that when there is ambiguity, we resolve the ambiguity beginning with the oldest (most basal) portions, amounting to an unacceptable assumption that the older the evolutionary event, the more accurate our information about it. There is always the possibility that an old event produced a terminal taxon (documented by ghost ranges/lineages in palaeontology). PACT analysis is, therefore, only the beginning of a study, not an end in itself.

Explaining the result of a PACT analysis begins a two-part process. First, we must distinguish general from unique nodes in an area cladogram. Unique nodes are those produced by an evolutionary event affecting only a single clade. At present, all

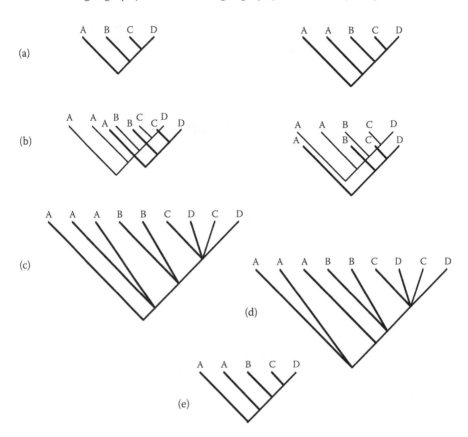

FIGURE 2.6   (a) Two taxon-area cladograms pertaining to areas A, B, C, and D. (b) The taxon-area cladograms aligned by common elements. There are two possible alignments, depending on the placement of area A in the taxon-area cladogram on the top left. (c) The taxon-area cladograms superimposed on each other. There are two possible alignments, depending on the placement of area A in the taxon-area cladogram on the top left. (d) PACT solution based on applying combination Rule 4 (Y(Y– + Y = (Y(Y–). Y(A)(Y– (A–) + Y(A–) = Y(A)(Y (A)–). There is a single PACT topology, (e) although there is still ambiguity about the placement of A in the taxon-area cladogram in the top left. This exemplar is also known as the 'problem of paralogy'.

nodes associated with evolutionary events affecting more than one clade are termed general nodes. Clearly, a general node involving 2 of 100 clades is not necessarily the same as one involving 75 of 100 clades. We do not have enough empirical data, or any models, to allow us to be more fine-grained in our distinctions amongst general nodes, and this is clearly an area of interest for future developments. Second, we must interpret the general nodes. Until recently, historical biogeographers believed that all general nodes in an area cladogram should be ascribed to vicariance. Lieberman (2000, 2003a, 2003b), however, noted that episodes of geodispersal (geographic areas fusing rather than splitting) could produce general nodes. In a similar vein, the *taxon pulse hypothesis* (Erwin, 1979, 1981) asserts that episodes of general biotic expansion ('jump dispersal') occurring between vicariance (or, more generally, isolation) events set the stage for future isolation events. Lieberman (2000, 2003a, 2003b) proposed a protocol for distinguishing general nodes in area cladograms due to isolation from those due to geodispersal, and Halas et al. (2005) showed that the same protocol could be used for cases of biotic expansion. The protocol is based on phylogenetic character optimisation (Lieberman used Fitch optimisation), using inhabited areas as the characters. General nodes associated with vicariance or other forms of isolation exhibit decreasing numbers of occupied areas relative to the next oldest node (Figure 2.7), whereas general nodes associated with

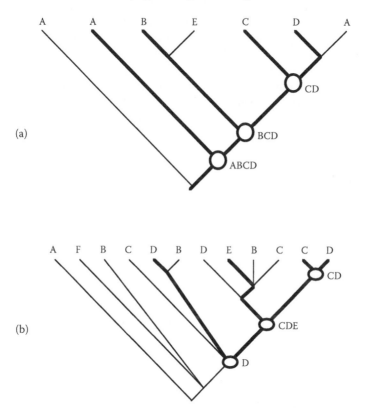

FIGURE 2.7 Nodal analysis for assessing the status of general nodes in area cladograms. (a) Thick branches indicate general nodes; thin branches represent elements of the GAC produced by only one clade. Progressive decrease in the number of areas associated with nodal values indicates that all general nodes are isolation nodes. This corresponds to either maximum vicariance or Hennig's progression rule. (b) Thick branches indicate general nodes; thin branches represent elements of the GAC produced by only one clade. Increase in the number of areas associated with nodal values from node 'D' to node 'CDE' indicates an episode of biotic expansion or general dispersal; decrease in the number of areas associated with nodal values from CDE to CD indicates an episode of isolation, corresponding to either vicariance or Hennig's progression rule. Alternation of biotic expansion and isolation nodes is characteristic of taxon pulse radiations.

biotic expansion (breakdown or crossing of a barrier) are associated with increasing numbers of occupied areas relative to the next oldest node. Nodes in which the area is the same as the ancestral one are designated within-area nodes.

2.3.2   AN EXAMPLE

Folinsbee and Brooks (2007) recently performed a PACT analysis that included living and fossil proboscideans (Shoshani, 1996), hyaenids (Werdelin and Solounias, 1991; Werdelin and Turner, 1996; Werdelin et al., 1994), and hominoids (Begun, 2001; Begun et al., 1997; Finarelli and Clyde, 2004) since the Miocene (Figure 2.8). These clades were each distributed worldwide during their evolutionary histories, but of course their geographic ranges have expanded and contracted repeatedly over time as species dispersed and became extinct. PACT analysis of the taxon-area cladograms derived from the phylogenies of the three clades produced the General Area Cladogram (GAC) in Figure 2.9. The GAC is a depiction of the most parsimonious means by which the clades could have dispersed, given both their phylogeny and geographic history. Areas are optimised using the two-pass protocol developed by Farris (1970), similar to ACCTRAN and DELTRAN optimisation of phylogenetic trees.

Of the 39 nodes on the GAC, 6 (or 15%) represent radiations of gomphotheres into North and South America that were not duplicated in either the hyaenid or hominoid clade. Of the other 34 nodes in the GAC, 17 (50%) are associated with simultaneous events in all three clades, an additional 9 (26%) are associated with events in two clades, and 8 (24%) are unique events (involving a single clade). The 26 nodes involving at least two of the three clades are designated the general nodes. Of these, 6 nodes are vicariance nodes, 9 are biotic expansion nodes, and 10 are speciation events that occur within a widespread area (within area nodes). The basal node is ambiguous.

Hominoids, hyaenids, and proboscideans have been geographically associated since at least the early Miocene, as evidenced by their shared presence at various fossil localities. The results of the PACT analysis do not conform to a simple vicariance scenario, or any other simple pattern of migration and speciation. Rather, these data suggest there were alternating episodes of vicariance and biotic expansion as well as unique speciation events in each of these three clades over the last 20 Ma, which supports the taxon pulse hypothesis, summarised in Figure 2.10 (redrawn and modified from Folinsbee and Brooks, 2007). Initially, all three clades are found in Africa; at node xx there is a vicariance event involving the hyaenids and proboscids; these two clades later undergo further dispersal and speciation together (nodes xix and xviii). Nodes xvii and xvi probably reflect two parts of the same biotic expansion event involving dispersal from Africa and subsequent speciation (Out of Africa 1). Between nodes xvii and xii in the middle of the GAC, there are four general biotic expansion nodes involving taxa expanding their ranges out of Africa and diversifying in Eurasia (Out of Africa 2). Node xii is a biotic expansion event out of Africa that begins the next sequence of speciation events. The GAC is optimised to Asia at this point, suggesting that during this time, most of the species of the three clades included in this analysis originated in Asia. Node x indicates a vicariant event

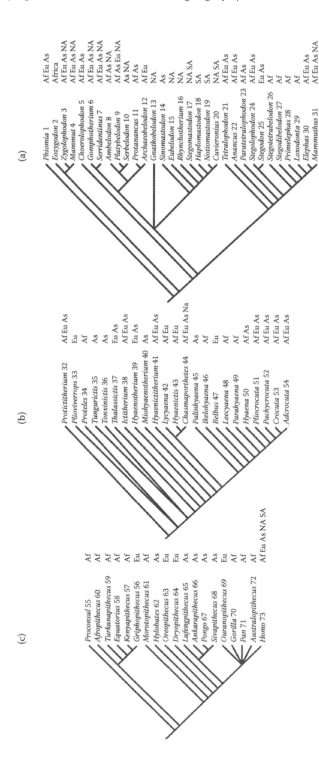

FIGURE 2.8 (a) Proboscidean taxon-area cladogram for PACT analysis (Shoshani, 1996). (b) Hyaenid taxon-area cladogram for PACT analysis (Werdelin and Solounias, 1991; Werdelin and Turner, 1996; Werdelin et al., 1994). (c) Hominoid taxon-area cladogram for PACT analysis (Begun et al., 1997; Finarelli and Clyde, 2004). Taxa are listed at the end of branches with number corresponding to placement on the general area cladogram (see Figure 2.9), and abbreviations of areas in which each taxon is found are placed after the taxon name (Af = Africa, Eu = Europe, As = Asia, NA = North America, SA = South America). (Redrawn and modified Folinsbee, K.E. and Brooks, D.R. (2007). *Journal of Biogeography*, 34: 383–397.)

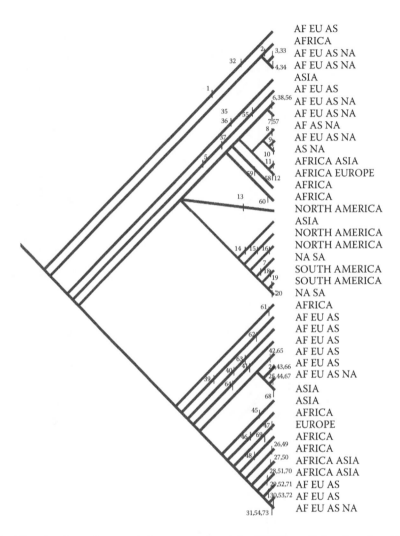

AF EU AS
AFRICA
AF EU AS NA
AF EU AS NA
ASIA
AF EU AS
AF EU AS NA
AF EU AS NA
AF AS NA
AF EU AS NA
AS NA
AFRICA ASIA
AFRICA EUROPE
AFRICA
AFRICA
NORTH AMERICA
ASIA
NORTH AMERICA
NORTH AMERICA
NA SA
SOUTH AMERICA
SOUTH AMERICA
NA SA
AFRICA
AF EU AS
AF EU AS
AF EU AS
AF EU AS
AF EU AS
AF EU AS NA
ASIA
ASIA
AFRICA
EUROPE
AFRICA
AFRICA
AFRICA ASIA
AFRICA ASIA
AF EU AS
AF EU AS
AF EU AS NA

FIGURE 2.9 General area cladogram produced by PACT analysis of proboscidean, hyaenid, and hominoid taxon-area cladograms (numbers and abbreviations as in Figure 2.8). (Redrawn and modified Folinsbee, K.E. and Brooks, D.R. (2007). *Journal of Biogeography*, 34: 383–397.)

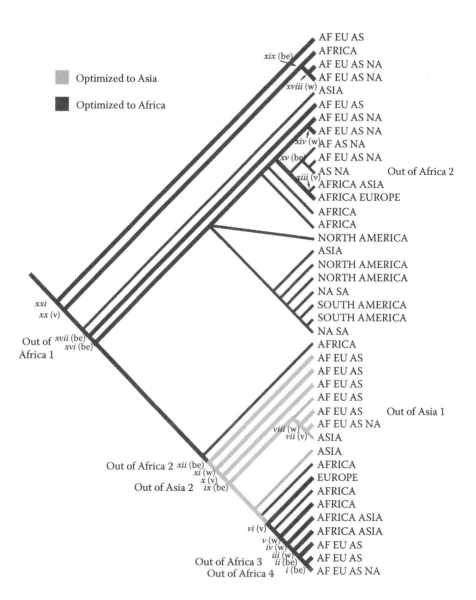

FIGURE 2.10 GAC with general nodes (ii, vii, etc.) resulting from vicariance (v), biotic expansion (be), and speciation events occurring within an ancestral area (w). Thick lines indicate branches on which two or more clades appear (general events) and thin lines reflect unique events affecting only a single clade. The determination of vicariance versus biotic expansion follows the protocol of Lieberman (2000, 2003a, 2003b). Major episodes of taxon pulse diversification are highlighted. See text for discussion. (Redrawn and modified Folinsbee, K.E. and Brooks, D.R. (2007). *Journal of Biogeography*, 34: 383–397.)

isolating Asia. This isolation event is followed by two parallel episodes of biotic expansion from Asia (Out of Asia 1 and 2), beginning at node ix. At node vi, there is another vicariant event, where most taxa go extinct in Eurasia and are therefore geographically restricted to Africa. At node v there is vicariant isolation in Africa followed by within area speciation (nodes iv and iii), then two more episodes of biotic expansion Out of Africa at nodes ii and i (Out of Africa 3 and 4).

Within this general pattern of taxon pulse-driven biotic evolution, each group also exhibits clade-specific histories of speciation and extinction. The Proboscidea underwent several radiations not paralleled by either the hominoids or hyaenids. The ambelodonts and gomphotheres (members of which persisted until the Pleistocene) evolved during one of these radiations beginning at node xvi and eventually occupied all five continents. The hyaenid input cladogram was highly pectinate (meaning there were few splitting events within the clade that produced more than a single genus). Fully half of the clade-specific events on the GAC (four of eight) occur in the hyaenids, so there is a high proportion of unique events in this clade. These are reconstructed as either peripheral isolates (if the descendant species is found outside the ancestral area) or within area speciation (if the descendant remains in the ancestral continent). The hominoids undergo an early clade-specific radiation in Africa (which produced early Miocene hominoid taxa) then track fairly closely with the other two clades during repeated expansion and contraction events.

Most evolutionary biologists and palaeontologists focus on a single particular clade or set of species to answer questions about modes of speciation and patterns of geographic distribution; this is due to operator choice rather than methodological blocks. That approach is limited by excluding data that might be pertinent to the set of species in which we are interested, because we know they did not evolve in a vacuum but must have interacted with unrelated taxa. In this example, hyaenas, hominoids, and proboscideans each have a unique and complex evolutionary history that includes migration, extinction, and radiation. Those historical patterns are intrinsically linked, occurring over the same landscape and under similar environmental and ecological conditions. PACT reconstructs historical patterns that are more complex than any of the individual clades involved in order to estimate a more complete picture of regional biogeography. A further strength of this new method is in its ability to combine data from both extant and extinct taxa, thereby illuminating patterns of evolutionary change.

2.3.3   Comparing What Is with What Ought to Be

The consilience perspective developed by Kant and Whewell serves us well in our transition from simplicity thinking to complexity thinking in historical biogeography. The more the results of empirical and theoretical work in historical biogeography converge, the closer we are to reconciling what is with what we think ought to be. This is true not just for historical biogeography, but for all types of comparative phylogenetic studies (Brooks and McLennan, 2002; Brooks and Van Veller, 2008), and even for phylogenetic analysis itself (Brooks et al., 2007).

If the world is not simple, our discovery methods must be as free of a priori prohibitions as possible. They must promise to find generalities in the data, that is, to point to a singular answer when the data support it. At the same time, they must promise not to give us a singular answer when the data are ambiguous. Until the advent of PACT, the only method of historical biogeography that permitted reticulated area relationships was secondary BPA (Brooks, 1990). Wojcicki and Brooks (2004, 2005) discovered that even secondary BPA had embedded assumptions about mechanisms that prejudged the data. PACT is free of those assumptions, but the trade-off is that there may be multiple interpretations of the taxon histories included in a general area cladogram produced by PACT.

DIVA (Ronquist, 1997) was an attempt to incorporate more than one phenomenon into historical biogeographic analyses, but it was not very successful; Brooks and McLennan (2002) showed that DIVA is insensitive to particular forms of dispersal, in addition to not producing reticulated area cladograms. Other authors have attempted to augment DIVA with additional information, particularly stratigraphic and palaeo-geographic data, in an effort to deal with complex datasets (Hunn and Upchurch, 2001; Meerow et al., 2006; Sanmartin, 2003; Sanmartin et al., 2007; Upchurch and Hunn, 2002; Upchurch et al., 2002). Nonetheless, all these studies followed traditional cladistic biogeographic practice in producing a simple, that is, nonreticulated, area cladogram presumed to represent the vicariance events, and then explaining nonfit-ting data in a variety of ways depending on the auxiliary data employed. Oberprieler (2005) used DIVA in investigating a presumptive reticulated set of area relationships, but was forced to stipulate the reticulated nature of the relationships prior to applying the method. As noted by Brooks and McLennan (2001), this is the potential flaw of model-based methods of analysis such as DIVA: they are exhaustive only with respect to the alternative hypotheses considered by the researcher. Since Wiley (1988a, 1988b) proposed Assumption 0 and Brooks (1990) proposed what is now called secondary BPA, there has been a mechanism for discovering not only whether there are reticu-lated area relationships in a dataset, but also for specifying which ones (Bouchard et al., 2004; Brooks, 1990, 1992, 2004, 2005; Brooks and Ferrao, 2005; Brooks and Folinsbee, 2005; e.g., Brooks and Hoberg, 2007; Brooks and McLennan, 1991, 1993a, 1993b, 1993c, 1994, 2001, 2002; Brooks and Van Veller, 2003, 2008; Folinsbee and Brooks, 2007; Green et al., 2002; Halas et al., 2005; Hoberg and Brooks, 2008; McLennan and Brooks, 2002; Spironello and Brooks, 2003; Van Veller and Brooks, 2001; Van Veller et al., 2003). We believe this explains why secondary BPA and PACT studies, rather than DIVA studies, were able to detect the signature of the taxon pulse and propose that it replace the vicariance model.

Although advocates of DIVA have not addressed the articles cited above, practitio-ners of DIVA seem to understand its shortcomings with respect to complex datasets. We believe this explains the modification of DIVA into TreeFitter, which emulates secondary BPA (e.g., Sanmartin et al., 2007). If TreeFitter emulates secondary BPA, it may contain the same flaws in secondary BPA that PACT corrects.

Once these potential problems are resolved, PACT results for a dataset could be used as the Assumption 0 null hypothesis for calibrating the costs of different events in TreeFitter. PACT produces general area cladograms that are not neces-sarily simple. With complexity, however, comes the possibility of ambiguity. For PACT analyses, just as with quantitative phylogenetic analysis, such ambiguity may produce multiple equally parsimonious GACs or multiple equally parsimonious placements of particular taxa in a GAC. Multiple equally parsimonious GACs generally result from insufficient data. Consider the following exemplar (with thanks to Paul Upchurch). The data consist of two taxon-area cladograms, (A(B(CD))) and (A(C(BD))). PACT produces four equally parsimonious GACs from these data, namely, (A(C(B(DC)))), (A(B(C(DB)))), (A(B(CBD))), and (A(C(BCD))). Based solely on these data, we cannot suggest a preference for one of these over the others. Brooks and McLennan (1991, 1993a) first noted that a minimum of three taxon-area cladograms were needed to be able to choose between the simplest of alternative patterns, that of choosing between explanations for the absence of a taxon in a particular area (i.e., was the taxon never in the area, or was it in the area and subsequently became extinct). Dubbed the 'Threes Rule' (Brooks and McLennan, 2002), this leads to a direct admonition to empiricists:

> When in doubt, get more data. (Brooks and McLennan, 2002)

Ambiguous results in placement of particular taxa in a GAC produced by PACT may arise from such things as polytomies in a taxon-area cladogram, which may indicate that the phylogenetic hypothesis is not well supported or that speciation at the polytomous node(s) did not leave a clear temporal and spatial signal, or from differential patterns of dispersal and differentiation amongst multiple clades in response to the same episode of biotic expansion (see Halas et al., 2005 for a discussion).

One way to assess the placement of ambiguous taxa on the GAC is to attempt to date each event in each taxon-area cladogram and then place ambiguous nodes and taxa in accordance with those estimated ages. At this point, we begin to enter a new level of consilience studies, one in which we look to minimise the difference between data and model, between reason and experience, and between what is and what ought to be. This is because our means of dating evolutionary events rely on inference, models of reason, rather than direct experience.

In the case of the study discussed above, we may use First Appearance Dates, which provide a *minimum* age for each node. The fossil record is somewhat limited, given that younger taxa are almost always going to be better represented than more ancient ones given the same environmental conditions (Kidwell and Holland, 2002). Although the fossil record is potentially incomplete, it provides data that are real. In general, the First Appearance Dates support the hypothesis that many of the general nodes were correlated events between at least two of these clades (Folinsbee and Brooks, 2007). Portions of the GAC which fit the stratigraphic record well indicate points at which reason and experience are in close accord; those portions which do not fit the taphonomic record well indicate cases in which we still have considerable work to do, empirical and theoretical, to narrow the gap between what is and what ought to be.

Researchers dealing with extant taxa often resort to using molecular clock estimates in lieu of stratigraphic data. Donoghue and Moore (2003) recently proposed that molecular clock estimates be used to place taxa at particular nodes in any area cladogram. Their concern was that independent evolutionary episodes involving the same areas could produce spuriously general patterns, a concern first articulated more than 20 years earlier by Endler (1982). This effort to avoid the pitfalls of one form of oversimplification, however, can open the door to another. Molecular clock estimates are often based on a particular gene, but the evolution of a gene is only part of the evolution of a species lineage. And the evolutionary clock of a species lineage must surely be some sort of statistical averaging over the clocks of each gene in the collective genome of the species (for a discussion, see Brooks and McLennan, 2002). Molecular clock estimates also use a fossil calibration point, which requires the same assumptions about the accuracy of stratigraphic data (Reisz and Muller, 2004).

A final way that the gulf between is and ought can be narrowed is the modification of our models of what ought to be. It is heartening to see event-based methods, such as DIVA, finally being modified along the lines initially suggested by Brooks (1990), to include complex area relationships (Sanmartin et al., 2008). Those models need to be extended further, allowing us to distinguish not between simple and complex historical biogeographic patterns, but also amongst different modes of evolutionary radiation leading to those patterns, beginning initially with ways to distinguish amongst vicariance, Hennig's progression rule, and the taxon pulse hypothesis.

2.4 CONCLUSIONS

It seems that adopting the most parsimonious inference allowed by the data is still the best we can do for empirical studies. And if the Kantian view is correct, and we cannot rely on reason or experience alone, we cannot assume that the world is simple. Therefore, we must use epistemological parsimony, not ontological parsimony, adopting working hypotheses that are as simple as the data allow, not as simple as we can imagine.

The results of biogeographic studies using PACT demonstrate that complex evolutionary and biogeographic patterns, such as those evidenced by a combination of independent phylogenies, may be best represented by a single general, yet complex, reconstruction. That we can extract general patterns from such disparate data does not imply that clade-specific exceptions to that general pattern do not exist. On the contrary, those exceptions do exist and can be discovered using the same methods that establish the general biogeographic relationships.

PACT adds two critical elements to studies of historical biogeography. First, it permits complex patterns to be found in complex data. Second, and following from the first, PACT facilitates the integration of fossil and recent taxa. If we look only at the geographic distributions and relationships of extant hominoids and their sister group, the cercopithecoids, the ancestral distribution would be interpreted as African, with movement to, and then diversification within, Asia (hylobatids and orangutans), followed by a return to Africa and additional diversification (gorillas, chimps, humans), with additional dispersal out of Africa (humans). Co-evolutionary analysis of some hominoid parasites, using PACT (Brooks and Ferrao, 2005), produced the same biogeographic patterns. In the study above, we showed that clades composed wholly or mostly of extinct taxa provide added depth and breadth to biogeographic analyses based only on extant taxa (Novacek, 1992), identifying additional general episodes and clade-specific events involving clades with which hominoids were associated ecologically. That study also adds palaeontological support to other recent studies that suggest the major driver of biotic diversification has been taxon pulses rather than simple vicariance (Bouchard et al., 2004; Halas et al., 2005; Spironello and Brooks, 2003).

All important theories, and the research programmes based on them, ensure their deaths by setting their explanatory goals too high. The maximum vicariance hypothesis was not stillborn: the earliest studies found general patterns that conformed to the hypothesis, and that was so unexpected that it was reasonable to believe that apparent exceptions to the simple general patterns were the result of incomplete knowledge, and therefore additional sampling would reinforce the general patterns. The hypothesis died of natural causes when increased knowledge led to overly complex manipulations of the data (various permutations of Assumptions 1 and 2, TAS, various event-based models, and now likelihood models). All those efforts have maintained the theory in a vegetative state, but it is brain dead. The maximum vicariance hypothesis has effectively been replaced by the taxon pulse model, so historical biogeography is not bereft of a robust null hypothesis.

SUMMARY

Historical contingency, an essential element of Darwinian evolution, suggests that biological evolution is complex. A major element of historical contingency is the interplay of organisms and geography, or historical biogeography. Historical biogeography benefited from, but was also constrained by, the phylogenetic systematics revolution. The search for general patterns used the maximum vicariance model, a simple view which did not allow reticulated area relationships, and which assumed that all general patterns were the result of isolation events. Newly proposed methods, such as PACT (Phylogenetic Analysis for Comparing Trees), capable of detecting complex biogeographic relationships, have allowed historical biogeographers to test that model. Analyses performed to date have shown that most areas have reticulated histories with respect to the species living in them, and general patterns may be due to isolation or biotic expansion events. The maximum vicariance model has been falsified, and its replacement is the taxon pulse hypothesis.

ACKNOWLEDGEMENTS

We express our thanks to the conference organisers for the invitation to present these ideas in oral and printed form. DRB acknowledges funding support from the Natural Sciences and Engineering Research Council (NSERC) of Canada.

LITERATURE CITED

Begun, D.R. (2001) African and Eurasian Miocene hominoids and the origins of the Hominidae. In *Phylogeny of the Neogene Hominoid Primates of Eurasia* (eds. L. de Bonis, G. Koufos, and P. Andrews), Cambridge University Press, Cambridge, UK, pp. 231–253.

Begun, D.R., Ward, C.V., and Rose, M.D. (1997) Events in hominoid evolution. In *Function, Phylogeny, and Fossils: Miocene Hominoid Evolution and Adaptations* (eds. D.R. Begun, C.V. Ward, and M.D. Rose), Plenum Press, New York, pp. 389–415.

Bouchard, P., Brooks, D.R., and Yeates, D.K. (2004) Mosaic macroevolution in Australian wet tropics arthropods: Community assemblage by taxon pulses. In *Rainforest: Past, Present, Future* (eds. C. Moritz and E. Bermingham), University of Chicago Press, Chicago, pp. 425–469.

Brooks, D.R. (1990) Parsimony analysis in historical biogeography and coevolution: Methodological and theoretical update. *Systematic Zoology,* 39: 14–30.

Brooks, D.R. (1992) Origins, diversification, and historical structure of the helminth fauna inhabiting neotropical fresh-water stingrays (Potamotrygonidae). *Journal of Parasitology,* 78: 588–595.

Brooks, D.R. (2004) Reticulations in historical biogeography: The triumph of time over space in evolution. In *Frontiers of Biogeography. New Directions in the Geography of Nature* (eds. M.V. Lomolino and L.R. Heaney), Sinauer, Sunderland, MA, pp. 125–144.

Brooks, D.R. (2005) Historical biogeography in the age of complexity: Expansion and integration. *Revista Mexicana de Biodiversidad,* 76: 79–94.

Brooks, D.R. and Ferrao, A. (2005) The historical biogeography of co-evolution: Emerging infectious diseases are evolutionary accidents waiting to happen. *Journal of Biogeography,* 32: 1291–1299.

Brooks, D.R. and Folinsbee, K.E. (2005) Paleobiogeography: Documenting the ebb and flow of evolutionary diversification. *Paleontological Society Papers,* 11: 15–43.

Brooks, D.R. and Hoberg, E.P. (2007) How will global climate change affect parasite-host assemblages? *Trends in Parasitology,* 23: 571–574.

Brooks, D.R. and McLennan, D.A. (1991) *Phylogeny, Ecology, and Behaviour: A Research Program in Comparative Biology,* University of Chicago Press, Chicago.

Brooks, D.R. and McLennan, D.A. (1993a) Historical ecology as a research program in macroevolution. In *Systematics, Historical Ecology, and North American Freshwater Fishes* (ed. R.L. Mayden), Stanford University Press, Stanford, CA, pp. 76–113.

Brooks, D.R. and McLennan, D.A. (1993b) Historical ecology: Examining phylogenetic components of community evolution. In *Species Diversity in Ecological Communities* (eds. R.E. Ricklefs and D. Schluter), University of Chicago Press, Chicago, pp. 267–280.

Brooks, D.R. and McLennan, D.A. (1993c) Macroevolutionary patterns of morphological diversification among parasitic flatworms (Platyhelminthes, Cercomeria). *Evolution,* 47: 495–509.

Brooks, D.R. and McLennan, D.A. (1994) Historical ecology as a research programme: Scope, limitations, and future. In *Phylogenetic Approaches to Ecology* (eds. P. Eggleton and R.I. Vane-Wright) Academic Press, London, pp. 1–27.

Brooks, D.R. and McLennan, D.A. (2001) A comparison of a discovery-based and an event-based method of historical biogeography. *Journal of Biogeography,* 28: 757–767.

Brooks, D.R. and McLennan, D.A. (2002) *The Nature of Diversity: An Evolutionary Voyage of Discovery,* University of Chicago Press, Chicago.

Brooks, D.R. and Van Veller, M.G.P. (2003) Critique of parsimony analysis of endemicity as a method of historical biogeography. *Journal of Biogeography,* 30: 819–825.

Brooks, D.R. and Van Veller, M.G.P. (2008) Assumption 0 analysis: Comparative phylogenetic studies in the age of complexity. *Annals of the Missouri Botanical Garden,* 95: 201–223.

Brooks, D.R., Bileweitch, J., Condy, C., Evans, D.C., Folinsbee, K.E., Fröbisch, J., Halas, D., Hill, S., McLennan, D.A., Mattern, M., Tsuji, L.A., Ward, J.L., Wahlberg, N., Zamparo, D., and Zanatta, D. (2007) Quantitative phylogenetic analysis in the 21st century. *Revista Mexicana de Biodiversidad,* 78: 225–252.

Darwin, C. (1872) *On the Origin of Species,* 6th edition, John Murray, London.

Depew, D.J. and Weber, B.H. (1995) *Darwinism Evolving: Systems Dynamics and the Genealogy of Natural Selection*, MIT Press, Cambridge, MA.

Donoghue, M.J. and Moore, B.R. (2003) Toward an integrative historical biogeography. *Integrative and Comparative Biology*, 43: 261–270.

Dowling, A.P.G. (2002) Testing the accuracy of TreeMap and Brooks parsimony analyses of coevolutionary patterns using artificial associations. *Cladistics*, 18: 416–435.

Dowling, A.P.G., van Veller, M.G.P., Hoberg, E.P., and Brooks, D.R. (2003) A priori and a posteriori methods in comparative evolutionary studies of host-parasite associations. *Cladistics*, 19: 240–253.

Ebach, M.C. and Morrone, J.J. (2005) Forum on historical biogeography: What is cladistic biogeography? *Journal of Biogeography*, 32: 2179–2187.

Endler, J.A. (1982) Problems in distinguishing historical from ecological factors in biogeography. *American Zoologist*, 22: 441–452.

Erwin, T.L. (1979) Thoughts on the evolutionary history of ground beetles: Hypotheses generated from comparative faunal analyses of lowland forest sites in temperate and tropical regions. In *Carabid Beetles—Their Evolution, Natural History, and Classification* (eds. T.L. Erwin, G.E. Ball, and D.R. Whitehead), W. Junk, The Hague, pp. 539–592.

Erwin, T.L. (1981) Taxon pulses, vicariance, and dispersal: An evolutionary synthesis illustrated by carabid beetles. In *Vicariance Biogeography—A Critique* (eds. G. Nelson and D.E. Rosen), Columbia University Press, New York, pp. 159–196.

Farris, J.S. (1970) Methods of computing Wagner trees. *Systematic Zoology*, 29: 83–92.

Finarelli, J.A. and Clyde, W.C. (2004) Reassessing hominoid phylogeny: Evaluating congruence in the morphological and temporal data. *Paleobiology*, 30: 614–651.

Folinsbee, K.E. and Brooks, D.R. (2007) Early hominoid biogeography: Pulses of dispersal and differentiation. *Journal of Biogeography*, 34: 383–397.

Gould, S.J. and Lewontin, R. (1979) The spandrels of San Marco and the Panglossian paradigm: A critique of the adaptationist programme. *Proceedings of the Royal Society of London*, B205: 581–598.

Green, M.D., Van Veller, M.G.P., and Brooks, D.R. (2002) Assessing modes of speciation: Range asymmetry and biogeographical congruence. *Cladistics*, 18: 112–124.

Halas, D., Zamparo, D., and Brooks, D.R. (2005) A historical biogeographical protocol for studying biotic diversification by taxon pulses. *Journal of Biogeography*, 32: 249–260.

Harvey, P.H. and Pagel, M.D. (1991) *The Comparative Method in Evolutionary Biology*, Oxford University Press, Oxford.

Hembree, D.I. (2006) Amphisbaenian paleobiogeography: Evidence of vicariance and geodispersal patterns. *Palaeogeography Palaeoclimatology Palaeoecology*, 235: 340–354.

Hennig, W. (1966) *Phylogenetic Systematics*, University of Illinois, Urbana.

Hoberg, E.P. and Brooks, D.R. (2008) A macroevolutionary mosaic: Episodic host-switching, geographical colonization and diversification in complex host-parasite systems. *Journal of Biogeography*, 35: 1533–1550.

Hunn, C.A. and Upchurch, P. (2001) The importance of time/space in diagnosing the causality of phylogenetic events: Towards a "chronobiogeographical" paradigm? *Systematic Biology*, 50: 1–17.

Kidwell, S.M. and Holland, S.M. (2002) The quality of the fossil record: Implications for evolutionary analysis. *Annual Review of Ecology and Systematics*, 33: 561–588.

Kluge, A.G. (1997a) Sophisticated falsification and research cycles: Consequences for differential character weighting in phylogenetic systematics. *Zoologica Scripta*, 26: 349–360.

Kluge, A.G. (1997b) Testability and the refutation and corroboration of cladistic hypotheses. *Cladistics*, 13: 81–96.

Lieberman, B. (2000) *Paleobiogeography*, Plenum/Kluwer Academic, New York.

Lieberman, B. (2003a) Paleobiogeography: The relevance of fossils to biogeography. *Annual Review of Ecology, Evolution, and Systematics*, 34: 51–69.

Lieberman, B. (2003b) Unifying theory and methodology in biogeography. *Evolutionary Biology,* 33: 1–25.

Lieberman, B. and Eldredge, N. (1996) Trilobite biogeography in the Middle Devonian: Geological processes and analytical methods. *Paleobiology,* 22: 66–79.

McLennan, D.A. and Brooks, D.R. (2002) Complex histories of speciation and dispersal: An example using some Australian birds. *Journal of Biogeography,* 29: 1055–1066.

Meerow, A.W., Francisco-Ortega, J., Kuhn, D.N., and Schnell, R.J. (2006) Phylogenetic relationships and biogeography within the Eurasian clade of Amaryllidaceae based on plastid ndhF and nrDNA ITS sequences: Lineage sorting in a reticulate area? *Systematic Botany,* 31: 42–60.

Neiman, S. (2002) *Evil in Modern Thought: An Alternative History of Philosophy.* Princeton University Press, Princeton, NJ.

Novacek, M.J. (1992) Fossils as critical data for phylogeny. In *Extinction and Phylogeny* (eds. M.J. Novacek and Q.D. Wheeler), Columbia University Press, New York, pp. 46–88.

Oberprieler, C. (2005) Temporal and spatial diversification of Circum-Mediterranean Compositae-Anthemideae. *Taxon,* 54: 951–966.

Reisz, R.R. and Muller, J. (2004) Molecular timescales and the fossil record: A paleontological perspective. *Trends in Genetics,* 20: 237–241.

Rode, A.L. and Lieberman, B.S. (2005a) Integrating evolution and biogeography: A case study involving Devonian Crustaceans. *Journal of Paleontology,* 79: 267–276.

Rode, A.L.S. and Lieberman, B.S. (2005b) Paleobiogeographic patterns in the Middle and Late Devonian emphasizing Laurentia. *Palaeogeography Palaeoclimatology Palaeoecology,* 222: 272–284.

Ronquist, F. (1997) Dispersal-vicariance analysis: A new approach to the quantification of historical biogeography. *Systematic Biology,* 46: 195–203.

Sanmartin, I. (2003) Dispersal vs. vicariance in the Mediterranean: Historical biogeography of the Palearctic Pachydeminae (Coleoptera, Scarabaeoidea). *Journal of Biogeography,* 30: 1883–1897.

Sanmartín, I., van der Mark, P., and Ronquist, F. (2008) Inferring dispersal: A Bayesian, phylogeny-based approach to island biogeography, with special reference to the Canary Islands. *Journal of Biogeography,* 35: 428–449.

Sanmartin, I., Wanntorp, L., and Winkworth, R.C. (2007) West wind drift revisited: Testing for directional dispersal in the southern hemisphere using event-based tree fitting. *Journal of Biogeography,* 34: 398–416.

Shoshani, J. (1996) Para- or monophyly of the gomphotheres and their position within Proboscidea. In *The Proboscidea: Evolution and Palaeoecology of the Elephants and Their Relatives* (eds. J. Shoshani and P. Tassy), Oxford University Press, Oxford, pp. 149–177.

Simpson, G.G. (1953) *The Major Features of Evolution,* Columbia University Press, New York.

Spironello, M. and Brooks, D.R. (2003) Dispersal and diversification in the evolution of Inseliellium, an archipelagic dipteran group. *Journal of Biogeography,* 30: 1563–1573.

Upchurch, P. and Hunn, C.A. (2002) 'Time'—The neglected dimension in cladistic biogeography. *Geobios,* 35: 277–286.

Upchurch, P., Hunn, C.A., and Norman, D.B. (2002) An analysis of dinosaurian biogeography: Evidence for the existence of vicariance and dispersal patterns caused by geological events. *Proceedings of the Royal Society of London Series* B269: 613–622.

Van Veller, M.G.P. and Brooks, D.R. (2001) When simplicity is not parsimonious: A priori and a posteriori methods in historical biogeography. *Journal of Biogeography,* 28: 1–11.

Van Veller, M.G.P., Brooks, D.R., and Zandee, M. (2003) Cladistic and phylogenetic biogeography: The art and the science of discovery. *Journal of Biogeography,* 30: 319–329.

Van Veller, M.G.P., Kornet, D.J., and Zandee, M. (2000) Methods in vicariance biogeography: Assessment of the implementations of assumptions 0, 1, and 2. *Cladistics,* 16: 319–345.

Van Veller, M.G.P., Kornet, D.J., and Zandee, M. (2002) A posteriori and a priori methodologies for testing hypotheses of causal processes in vicariance biogeography. *Cladistics,* 18: 207–217.

Van Veller, M.G.P., Zandee, M., and Kornet, D.J. (1999) Two requirements for obtaining valid common patterns under different assumptions in vicariance biogeography. *Cladistics,* 15: 393–406.

Werdelin, L. and Solounias, N. (1991) The Hyaenidae: Taxonomy, systematics, and evolution. *Fossils and Strata,* 30: 1–104.

Werdelin, L. and Turner, A. (1996) The fossil and living Hyaenidae of Africa: Present status. In *Palaeoecology and Palaeoenvironments of Late Cenozoic Mammals: Tributes to the Career of C.S. (Rufus) Churcher* (eds. K.M. Stewart and K.L. Seymour), University of Toronto Press, Toronto, pp. 637–659.

Werdelin, L., Turner, A., and Solounias, N. (1994) Studies of fossil hyaenids: The genera *Hyaenictis* Gaudry and *Chasmoporthetes* Hay, with a reconstruction of the Hyaenidae of Langebaanweg, South Africa. *Zoological Journal of the Linnean Society,* 111: 197–217.

Whewell, W. (1858) *The History of Scientific Ideas.* John W. Parker, London.

Wiley, E.O. (1981) *Phylogenetics: The Theory and Practice of Phylogenetic Systematics,* Wiley InterScience, New York.

Wiley, E.O. (1986) Methods in vicariance biogeography. In *Systematics and Evolution* (ed. P. Hovenkamp), University of Utrecht Press, Utrecht, pp. 233–247.

Wiley, E.O. (1988a) Parsimony analysis and vicariance biogeography. *Systematic Zoology,* 37: 271–290.

Wiley, E.O. (1988b) Vicariance biogeography. *Annual Review of Ecology and Systematics,* 19: 513–542.

Wojcicki, M. and Brooks, D.R. (2004) Escaping the matrix: A new algorithm for phylogenetic comparative studies of co-evolution. *Cladistics,* 20: 341–361.

Wojcicki, M. and Brooks, D.R. (2005) PACT: An efficient and powerful algorithm for generating area cladograms. *Journal of Biogeography,* 32: 755–774.

Zandee, M. and Roos, M. (1987) Component compatibility in historical biogeography. *Cladistics,* 3: 305–332.

3 Uncertainties in Phanerozoic Global Continental Reconstructions and Their Biogeographical Implications
A Review

Alan G. Smith

CONTENTS

3.1   INTRODUCTION

The need for global reconstructions for biological research stems from the fact that the Earth's outer layer is broken up into tectonic plates that move relative to one another. As a result, its geography changes dramatically in geological time. For a given plate some of the geographical changes are quite rapid; for example, coast-lines may vary significantly as a result of the melting of ice sheets on a timescale of $\sim 10^4$–10^6 years. Longer-term geographical changes are brought about by large-scale uplift and subsidence on timescales of 10^6–10^7 years. However, the number and dis-tribution of the plates themselves varies significantly on timescales of 10^8 years or more. All of these changes have profound effects on evolution because they bring about vicariant events (i.e., events that separate or divide a group of organisms by a geographical barrier), resulting in differentiation of the original group into new varieties or species.

The continental parts of plates act as gigantic rafts that ferry shallow marine and terrestrial life into different latitudes and, during continental collisions, can cause major faunal and floral mixing. Similarly, for the oceans there are periods in Earth history when they have been freely connected to one another, and others when continents block exchanges of floras and faunas between neighbouring oceans, as in the case of the Americas today, where interchange between Pacific and Atlantic marine life is now prevented by the Isthmus of Panama. The transitions between a blocked and a free ocean circulation take place via the opening of *oceanic gateways*, separating two or more continents. Opening leads to isolation of terrestrial life on the affected continents. Closure of oceanic gateways has the opposite effect. Opening and closing of gateways may also have had a profound effect on climate, which in turn will add to evolutionary pressures (Pickering and Smith, 2001; Smith and Pickering, 2003).

The principles used in repositioning the continents have been known for many years (e.g., Smith et al., 1973). What is new, apart from a deeper understanding of how the Earth works as a planet, is the availability of fast, low-cost, high-storage computers and of civilian and declassified military datasets that show, amongst other things, the Earth's topography and bathymetry in amazing detail. The datasets are available in either gridded or vector format, and can be converted from one format to the other by standard geographical information systems.

This chapter discusses how continents may be repositioned in the past, and the errors involved in determining those past positions. It outlines the uncertainties in the shapes of continents and their former coastlines, what causes these uncertainties, and suggests how these may be decreased. The resulting continental reconstructions reduce one of the major uncertainties in understanding palaeobiogeographical distributions, but there are several important problems that cannot be solved by the available geological data, and possibly never will be. In particular, the errors in plate tectonic reconstructions do not allow a precise determination of the times when oceanic gateways opened to marine faunal interchange or closed to allow terrestrial faunal interchange: these times are much better defined by the palaeobiological data, although such data are not always available.

The basic architecture of the Earth is much better known than it was a third of a century ago, and its origins are much better understood and well described in standard texts. Only the outlines of this basic information are given here; in all cases the reader is referred to books such as Fowler (2005) for a more detailed treatment.

3.2 ABBREVIATIONS AND DEFINITIONS

λ (lambda) is the latitude or palaeolatitude of a point on the Earth's surface.

ϕ (phi) is the longitude or palaeolongitude of a point on the Earth's surface.

α_{95} (alpha ninety-five) is the radius in degrees of a circle of 95% confidence, which is centred on the mean pole of a set of palaeomagnetic poles, within which the true mean direction lies.

D is the magnetic declination, or the difference in degrees between geographic north and the direction of the horizontal component of magnetisation in a rock.

I is the magnetic inclination, or the angle (= dip) that the magnetisation in a rock makes with the horizontal.

GADF is the geocentric axisymmetric dipole field. This is the field that would be made by a bar magnet at the Earth's centre aligned along the Earth's geographic (= spin) axis.

APWP is the Apparent Polar Wander Path. The APWP is the trace of the mean palaeomagnetic poles of different ages relative to a continent, which is taken as fixed.

Gondwana is the name of the major Cambrian–Carboniferous continent composed of South America, Africa, Arabia, Madagascar, Antarctica, India, and Australia.

Laurentia is the name of the late Precambrian–Ordovician continent composed mostly of North America, Greenland, and northwest Scotland.

Laurussia is the name of the Silurian–Carboniferous continent composed of Laurentia and Baltica.

Pangaea is the large Permian to Triassic continent made by the amalgamation of Gondwana and Laurussia. Pangaea-A is the Gondwana–Laurussia reassembly made by fitting the continental margin of eastern North America to that of northwest Africa.

Pangaea-B is the Gondwana–Laurussia reassembly made by fitting the mean Permian palaeomagnetic poles of Gondwana and Laurussia together and simultaneously bringing the two major continents into contact.

LIP (Large Igneous Province) is a large basaltic continental or oceanic igneous province, generally composed at the surface of 'flood basalts' (e.g., Deccan Traps, Iceland).

3.3   TECTONIC PLATES

Plates make up the Earth's lithosphere, a rigid, mechanically strong, outer layer that is about 100 km thick, although it is very thin at spreading oceanic ridges and may be thicker under large areas of older Precambrian rocks (the 'shields' or 'cratons'). The name lithosphere comes from the Greek *lithos* = rock. Below it is a much weaker layer known as the *asthenosphere* (from the Greek *asthenia* = weak).

Earthquakes are caused by the catastrophic release of elastic energy built up in brittle rocks by their steady deformation. The general lack of seismicity within plates means that these areas are not being deformed in the upper brittle layer (i.e., they are rigid). The types of motion that can occur between rigid areas on a sphere are: (a) sliding past one another, giving rise to translation; (b) moving apart, giving rise to extension; and (c) moving toward each other, giving rise to convergence.

At the surface, active seismicity marks out the plate boundaries (Figure 3.1). The seismic zones divide the Earth's surface into areas with few or no earthquakes, which are known as tectonic plates. Present-day plates have irregular shapes, sizes, and components. For many purposes only 13 plates are required to give an adequate description of the plate motions (Figure 3.1), but up to 52 plates have been recognised (Bird, 2003). Some plates, such as the Cocos plate off western South America, are made entirely of oceanic lithosphere. The largest, the Pacific plate, is entirely oceanic lithosphere except for the small continental fragment of southern California, not visible on Figure 3.1. None of the 13 plates is made up only of continental lithosphere; all have some oceanic lithosphere attached to them: for example, the African

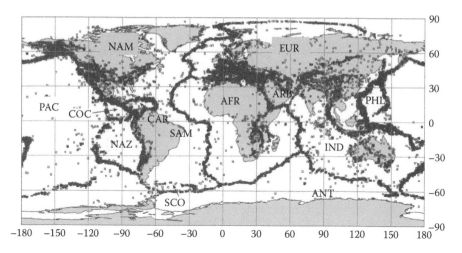

FIGURE 3.1 Earthquakes in depth range 0–25 km (open squares) mark out the plate boundaries. 13 plates are shown: AFR = Africa; ARB = Arabia; ANT = Antarctica; CAR = Caribbean; COC = Cocos; EUR = Eurasia; IND = India; NAM = North America; NAZ = Nazca; PAC = Pacific; PHL = Philippines; SAM = South America; SCO = Scotia. On this map the Indian plate includes Australia and is sometimes referred to as the Australian plate or the Indo-Australian plate. At the present time, there is a band of seismicity crossing the plate just south of India, which may mark the division between an Indian plate to the north and an Australian plate to the south, though at least one additional plate has also been proposed. The location of the tiny Juan de Fuca plate off western North America is obliterated by the many earthquakes in the region.

plate has a significant amount of the Atlantic and Indian Ocean lithosphere attached to it.

3.3.1 MOVING RIGID OBJECTS ON A SPHERE

All continents contain a region—which will be large for a major continent—that is essentially rigid when looked at on a global scale. How a rigid body moves on a sphere can be described mathematically by a theorem known as Euler's fixed point theorem (Bullard et al., 1965). If the rigid body starts in one position and moves to a new position, then the new position is reached by turning the body about a line that passes through the centre of the sphere through the *Euler angle*, θ (Figure 3.2a). The line cuts the sphere's surface at P, the *Euler pole*, which has a latitude, λ, and longitude, ϕ. The *Euler rotation* is given as (λ, ϕ, θ).

If two rigid bodies slide past each other, their boundary must lie on a small circle (i.e., a latitude line) with respect to the Euler pole (Figure 3.2b). All other motions cause the bodies to separate or converge.

Euler rotations can therefore be used to describe how the rigid parts of continents and ocean floor have moved in time. Geophysical data allow Euler rotations for oceans and continents to be estimated from the geological record, principally from

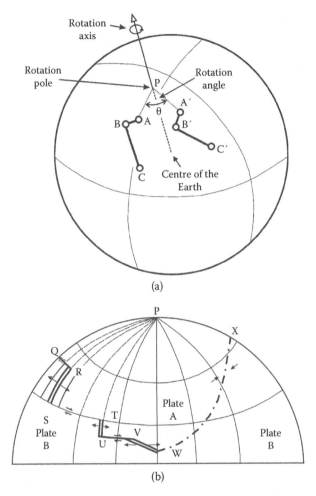

FIGURE 3.2 (a) The line ABC on a rigid body on the surface of a sphere (e.g. Earth) is moved to A′B′C′ by a rotation through the angle θ, about the rotation axis. The rotation passes through the centre of the Earth and cuts the surface at point P with geographic coordinates of latitude λ and longitude φ (not shown), leading to an Euler rotation of (λ, φ, θ). (b) An irregular plate margin separates plates A and B. The diagram shows the Euler pole, P, at the top of the figure. Any movement along latitude lines relative to the pole causes A and B to slide past one another. In the diagram, sliding (transform faulting) takes place along segments QR, ST and UV. The direction of extension is along latitude lines. Extension at right angles to ST takes place along RS and TU; with oblique extension along VW. WX is a convergent margin, with the convergence direction being along latitude lines. A low angle of convergence takes place on the plate margin near W, with less oblique convergence as one moves toward X.

ocean floor ridge-and-transform systems, best fits of continental margins, palaeo-magnetic data, and hot-spots, from all of which global maps can be made.

3.3.2 Oceanic Ridge-and-Transform Systems

3.3.2.1 Geometry

As continents separate from one another, the intervening ocean floor grows between them. The width of the seismic belt in a spreading ocean ridge system is very small (i.e., the active plate margin is a narrow zone and can be pinned down very precisely; Figure 3.1). In detail, the bathymetry probably provides even better locations of the ridge system than earthquakes.

Most ridge systems are made up of spreading ridge segments (extensional plate boundaries) and transform fault segments (conservative plate boundaries) that connect ridge segments together, generally at right angles to the spreading segments (Figures 3.2b and 3.3). The reasons for this unexpected and almost universal geometry are unclear. A few ridges, such as the Reykjanes Ridge southwest of Iceland, have ridge segments that are not at right angles to the transform segments: the ridge

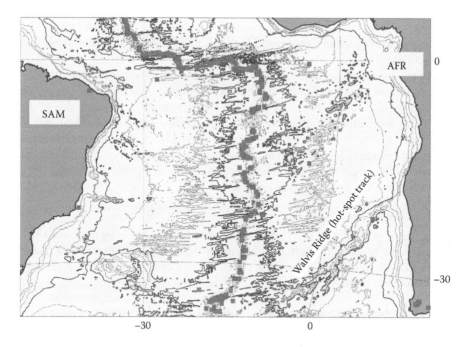

FIGURE 3.3 Map of the South Atlantic between South America (SAM) and Africa (AFR) showing earthquakes 0–25 km (filled squares) and submarine bathymetry at 1 km depth intervals. The 4 km contour is in a bolder line. The seismic activity shows a rectilinear pattern, particularly in the north, with north-trending segments marking the ridge segments and the E-trending segments marking transform faults. Inactive transform, or fracture zones, give a strong E-W grain to the topography. Cutting obliquely across this ocean-floor topography is the trace of the Walvis Ridge, a hot-spot track. Its conjugate to the W is less obvious.

segments must be separating obliquely to the transforms, analogous to segment VW in Figure 3.2b. Others do not join together, but form 'propagating rifts' (not shown) as in the Easter Island Ridge (e.g., Rusby and Searle, 1995). Apart from these exceptions, ocean ridge systems are, on a global scale, close to ideal plate boundaries: negligible in width and readily located.

Although the process of creating the ocean floor is widely known as seafloor spreading, it is referred to here as ocean floor spreading because it is an oceanic, rather than just a marine, process (e.g., the North Sea was not created by seafloor spreading); and it also highlights the contrast between oceans and continents, which is the most fundamental subdivision in tectonics.

3.3.2.2   Magnetic Polarity Timescale

Oceanfloor is generated at active ridge segments. As the new floor cools, it takes on the magnetic field of the time. Reversals of the Earth's magnetic field are randomly distributed in time but on average take place roughly every million years or so (Figure 3.4). When a polarity change takes place, the new ocean floor takes on the new polarity. With the passage of time the pattern of polarity changes shows up on the ocean floor as magnetic stripes of higher-than-average magnetisation alternating with stripes of lower-than-average magnetisation. The stripes impose a sort of magnetic bar code on the ocean floor that is unique and enables the polarity transitions to be identified (e.g., McElhinny and MacFadden, 2000) and dated (e.g., Gradstein et al., 2004).

The pattern on the ocean floor is generally remarkably symmetric. For example, if the pattern on one side of a mid-ocean ridge is reflected in an imaginary mirror on the ridge itself, the result is a close match to the pattern on the opposite side of the ridge. This symmetry is a reflection of the fact that the ridge is the weakest part of the ocean floor and breaks along it, rather than elsewhere, each time extension takes place. The symmetry is not perfect (e.g., between Australia and Antarctica; Müller et al., 2000); other areas show discontinuities in the spreading pattern, when the ridge appears to have jumped to another location.

3.3.2.3   Flow Lines

The actual flow lines that mark out how the ocean floor has grown between two continents are revealed by bathymetric maps of the ocean floor (Figure 3.3). These clearly show the positions of transform faults, and of former, now inactive, transform faults known as *fracture zones*.

3.3.2.4   Euler Rotations from Flow Lines and Magnetic Anomaly Patterns

As discussed above, rotations about an Euler pole give rise to small circle arcs, with the pole as their centre (Figure 3.1). A given flow line in the ocean floor approximates closely to a series of small circle arcs representing successive changes in the position of the Euler pole as the ocean floor grows. Thus one of the stages can be taken, its arc found, and an estimate obtained for the Euler pole for that stage. By using all flow lines showing arcs for that stage, one can find the best-fit estimate for the position of the Euler pole. The angle for that stage is given, as either the best-fit angle between the ends of the arcs, or the difference between the best match for the

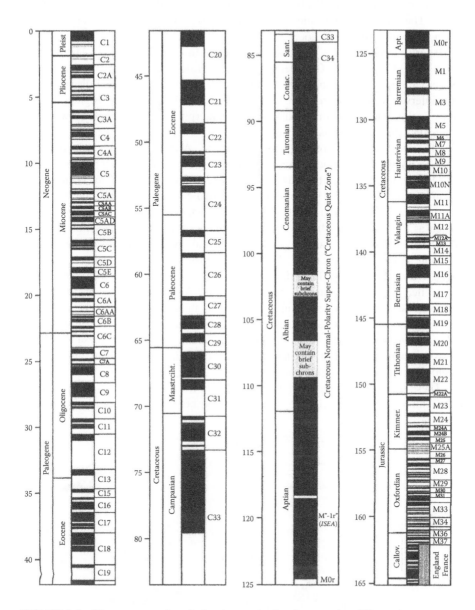

FIGURE 3.4 The geomagnetic polarity time scale showing ages in millions of years (Ma) on the left; followed by stratigraphic names, polarities and polarity nomenclature, modified from Fowler (2005, Figure 2.14). Black boxes are normal polarity intervals; white are reversed intervals; grey boxes are of uncertain polarity. The Quaternary and Tertiary (not shown) correspond approximately to the Pleistocene, and to the Palaeocene to Pliocene intervals respectively.

polarity position marking the beginning of the stage and that marking the end of the stage. The Euler pole and Euler angle together give the Euler rotation for that stage in the growth of the ocean floor.

Polarity transitions are irregularly distributed in time (Figure 3.4). The interval from ~125–84 Ma in the Cretaceous period, the Long Normal-Polarity Cretaceous Super-Chron, or 'Cretaceous Quiet Zone' is one of normal polarity. Euler poles for this interval are known from the flow line data, but the age to be assigned to each stage in this zone is generally found by linear interpolation of the angle between the anomalies that bracket this interval.

The resulting stage rotations for the Arctic, Atlantic, Indian, and Antarctic Oceans allow all the rigid areas of the major continents to be repositioned relative to one another as far back as the oldest ocean floor that separates each pair of continents (Figure 3.5). The positions of the two continents at the time when the ocean first started spreading is the best-fit of the edges of the continents, commonly taken as 2–4 km depth. In this chapter the edge of a continent is set at 2,000-m depth (see below). The transition from continental crust to oceanic crust is highly variable and usually takes place via a zone of transitional crust, probably best envisaged as a zone of extended and thinned continental crust peppered with structurally controlled intrusions. Examples include northwest Australia (Robb et al., 2005) and Norway (Scheck-Wenderroth et al., 2007). Some margins are asymmetric, as in the Labrador Sea (Srivastava and Roest, 1999), which has a very thin transitional crust of unknown nature that is 160-km wide (Chian et al., 1995). All of these variations give rise to problems when making detailed continental reconstructions, although on a global scale none are significant.

These rotations provide the fundamental data for making high-precision global reassemblies of the rigid continental areas for most of Mesozoic and all of Cenozoic

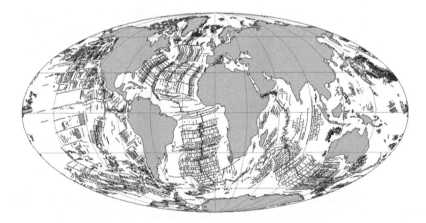

FIGURE 3.5 Observed magnetic anomalies and fracture zones from NOAA (National Oceanic and Atmospheric Administration). The observed anomalies can be extrapolated to make 'synthetic anomalies' that cover most of the oceans, except for parts of the Pacific and Indian Oceans. The broad white strip in the South Atlantic close to the west African and eastern South American margins reflects the absence of polarity changes during the Cretaceous Normal-Polarity Superchron (Figure 3.4).

time. A reassembly is not a map because it does not show (palaeo)latitude and (palaeo) longitude: any continent can act as a reference for a reassembly. To turn a reassembly into a map we need data from palaeomagnetism or hot-spots.

3.3.3 PALAEOMAGNETIC DATA

3.3.3.1 Declination and Inclination

The orientation of the magnetic field on the Earth's surface and in rocks, is described by its *declination, D,* or the bearing (azimuth), relative to geographical north, and the *inclination, I,* or 'dip' of the field (Figure 3.6). The field lines have a directional sense, which can be up or down relative to the surface. D, I, and the sense of the field uniquely specify the field at a point and are independent of any interpretation of it.

3.3.3.2 Rock Magnetism

Most rocks contain sufficient magnetic minerals, such as magnetite or haematite, for their D and I to be measured. If the orientation of a sample is noted when it is collected, then the direction of the field measured in the laboratory can be turned into its actual orientation in the field. In practice, the observed magnetism may be the sum of several different magnetic components of different minerals within the rock that are of different ages. From the point of view of making global reconstructions, one needs the D and I of the field at the time the rock was formed. Removing the effects of later, or secondary, magnetisation from the primary magnetisation is complex (McElhinny and Macfadden, 2000). Once the primary component has been isolated, it needs to be dated. It is generally assumed that if there are no obvious

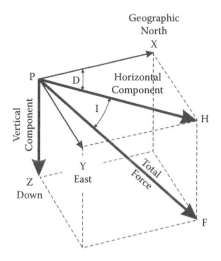

FIGURE 3.6   Declination, D, ('dip') and inclination, I, (bearing or azimuth) are the fundamental observations needed to record the direction of the field at any site. The field also has a sense, i.e. upward pointing, or downward pointing. The convention is to take field lines north of the magnetic equator as downward pointing (Figure 3.7).

signs of subsequent alteration to the minerals holding the magnetism, the age of the magnetisation is the age of the rock.

3.3.3.3 Earth's Magnetic Field

Many detailed measurements of the direction of the magnetic field in rocks in the age range 0–1 Ma show that the average field approximates closely to the field of a dipole (bar magnet) at the Earth's centre, whose axis parallels the spin axis (Figure 3.7). This relationship is assumed to have been applicable in the past and gives rise to the basic assumption of all palaeomagnetic interpretations: the Earth's field is geocentric (i.e., the bar magnet is not displaced along the spin [= geographical] axis); it is axisymmetric (i.e., looked at from above the North or South Pole, there is radial symmetry); and it is a dipole. All of these properties are those of a Geocentric Axisymmetric Dipole Field, or GADF (Figure 3.7).

For a dipole, a very simple expression relates the inclination to palaeolatitude:

$$\tan (\text{inclination}) = 2 \tan (\text{palaeolatitude})$$

Currently, the Earth's field deviates from the GADF model. The axis of the best-fitting dipole is inclined at about 11° to the Earth's spin axis because of the presence of nondipole components in the field (i.e., those parts of the field that do not fit the dipole model). The variations in the nondipole components give rise to secular variation, but when averaged over ~10^4 years or so, the secular variation averages to zero, and the mean magnetic pole of the Earth then averages to the Earth's spin axis (McElhinny and Macfadden, 2000).

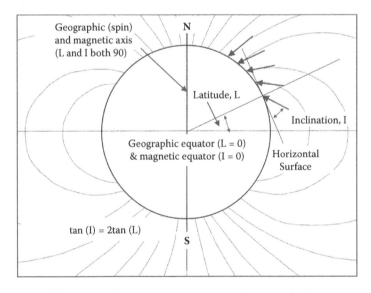

FIGURE 3.7 The GADF (Geocentric Axisymmetric Dipole Field) model fundamental to palaeomagnetic work. For such a model tan (I) = 2 tan (latitude).

3.3.3.4   The Palaeomagnetic Database

The International Association of Geomagnetism and Aeronomy maintains links to a database of palaeomagnetic data that can be downloaded from the web. The database contains nearly all palaeomagnetic measurements, together with details on the methods used, stratigraphic ages, and other relevant data. Thus poles that are suitable for orienting fragments are included along with poles that are unsuitable: the user must select those poles that are suitable for making reconstructions by using certain selection criteria. Commonly used criteria are given by Van der Voo (1988, 1990). Differences in the selection criteria, and therefore in the poles considered suitable, are one of the sources of the major differences between different pre-Mesozoic global reconstructions.

As noted above, the GADF model is supported by a plot of all poles in the database in the age range 0–1 Ma. During this interval plate motions are negligible (1° at most). There are 210 poles, only two of which lie at more than 45° from the pole and probably represent a transient field during a polarity transition. The mean pole of all these data is at latitude of 88.9°. The circular error, expressed as α_{95}, is 0.9°. α_{95} is simply the radius of a circle on the Earth's surface centred on the mean pole within which there is a 95% probability that the true mean pole lies (McElhinny and Macfadden, 2000). Thus, when errors are included, the mean magnetic pole for the past 1 m.y. lies less than 0.2° from the geographical pole and is essentially indistinguishable from it (Figure 3.8).

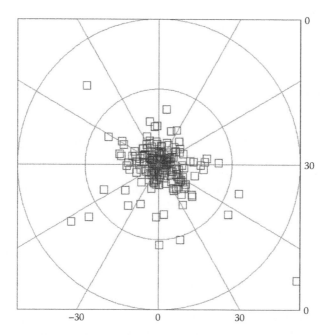

FIGURE 3.8   The 210 north poles for 1–0 Ma cluster around the present geographic North Pole with a mean latitude of 88.9°.

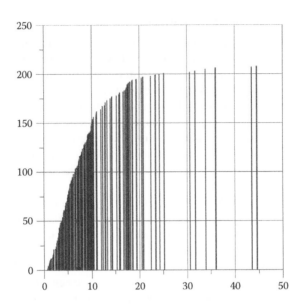

FIGURE 3.9   A bar chart showing the variation of the angular distance from the pole for 210 poles in the range 0–1 Ma. 2 poles that are more than 45° from the mean pole have been omitted. More than 96% of the poles are less than 25° from the mean pole.

A graph showing the angular distance against sample number shows that only 7 of the 210 poles, ~3%, lie more than 25° from the pole (Figure 3.9). It is assumed that a similar relationship held in past time, and that a similar 25° cut-off can be applied to older data.

To find a mean pole for a reassembly one selects all the 'reliable' North (or South) Poles for that fragment which spans the age of the map, for example, 10 Ma ± 1 m.y. for a 10 Ma map, or perhaps 100 Ma ± 10 m.y. for a 100 Ma map, and so on, then rotates the selected poles to a reference fragment such as Africa. The mean of the selected rotated poles on the reference fragment is the mean pole for the whole reassembly. Obviously, the same procedure can be applied to individual continents.

In this chapter all poles whose magnetisation age is not the same as the rock age are rejected. However, strict application of additional selection criteria (Van der Voo, 1988, 1990), such as the application of certain magnetic tests, results in too few poles for making reconstructions.

There are several ways of deriving a mean pole from a dataset. A commonly used method is to place a stratigraphic window (such as Jurassic) or an age range (such as 100–60 Ma) on the data, and calculate its α_{95}. The window is then moved to, for example, Cretaceous or 80–40 Ma and the calculation repeated (Figure 3.10). A newer method fits a cubic spherical spline to the data (Jupp and Kent, 1987) with the relative reliability of the mean poles being determined by other spline methods (Torsvik et al., 1992). The main advantage of a sliding window (or analysis period by period) is that it gives a numerical value for the error of each mean pole, whereas the spline method gives only a relative value. The main disadvantage is that, unlike

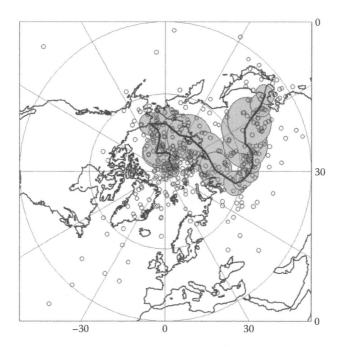

FIGURE 3.10 Azimuthal equal-area projection of the palaeomagnetic poles from North America in the age range 320–0 Ma for which the magnetization age is also the age of the rock. The large scatter can be reduced by eliminating poles that do not meet additional selection criteria (e.g. Van der Voo, 1988, 1990). However, this does not significantly alter the resulting apparent polar wander path (APWP), except to reduce the uncertainty. The APWP is shown as a thick black line. It has been calculated at steps of 10 m.y., with a sliding window of 40 m.y. All poles that lie more than 30° from the mean pole for each age step have been removed. The result is similar to that of McElhinny and McFadden (2000, Table 6.5, p. 254), although they used stratigraphic periods, rather than equal time steps to determine the result. Grey circles are the α-95 circles that give a measure of the uncertainty in mean pole.

a fitted spline, the resulting APWP is not smooth but is a zigzag made of great circle arcs of differing lengths. Also, the spline method allows poles to have different weights, whereas for the estimation of α_{95}, each pole is given the same weight.

In this chapter, the traditional method is used, with a sliding window of 40 m.y. and rejection of all poles >25° from the mean pole, a value suggested by the distribution of poles for the past 1 m.y. discussed above (Figure 3.8). The resulting α_{95}s are probably minimum values because the selection excludes those poles that lie beyond the 25° limit that otherwise meet the palaeomagnetic criteria.

The range of α_{95} values is shown for four datasets (Figure 3.11): (1) all the stable continental areas for 240–0 Ma, (2) Phanerozoic Gondwana, (3) Phanerozoic Laurentia, and (4) Phanerozoic Siberia. The mean global α_{95} values are surprisingly small: less than 2° for 240–0 Ma. For Laurentia and Gondwana α_{95} is generally less than 4°, and increases with age. For Siberia, the α_{95}s are much larger, with four windows for which $\alpha_{95} > 8°$, including one at 200 Ma, which is >25°.

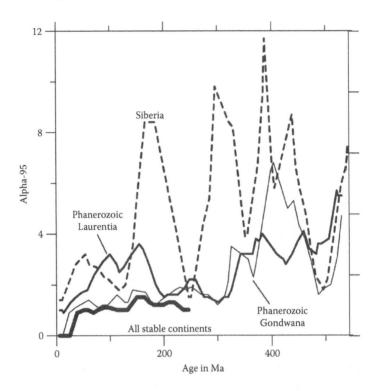

FIGURE 3.11   The range of uncertainty in mean palaeomagnetic pole positions for poles that satisfy minimum selection criteria (see text). The thick short curve is for 240–0 Ma for poles from the tectonically stable parts of the major continents. The less thick long line is for Phanerozoic poles from Laurentia only (North America and Greenland). The thin line is for Phanerozoic poles for the Gondwana continents. The dashed line is for Phanerozoic poles from Siberia; α_{95} for Siberian poles in the window 220–180 Ma have been omitted because $\alpha_{95} > 25°$. In general, the higher the α_{95}, the smaller the number of poles.

3.3.3.5   Turning a Reassembly into a Map

A North (or South) palaeomagnetic mean pole gives an estimate of the position of the North (or South) geographical pole of the time. The palaeomagnetic pole may be placed at the geographical pole by applying the appropriate rotation. For example, if a palaeomagnetic North pole is at a latitude λ and a longitude ϕ, then the Euler rotation (Euler latitude, Euler longitude, Euler angle) that brings the palaeomagnetic pole into coincidence with the geographical pole is $(0, \phi-90, 90-\lambda)$. The Euler pole is on the equator. For a South pole, the rotation is $(0, \phi-90, \lambda-90)$. These rotations are used to move a fragment into palaeogeographic co-ordinates.

3.4   HOT-SPOTS AND THEIR MOTIONS

3.4.1   HOT-SPOTS

Hot-spots are areas where there has been intense basaltic magmatism (see below). A nongenetic name for flood basalts on continents and for oceanic plateaus, which are their oceanic equivalents, is 'Large Igneous Provinces' or LIPs (Coffin and Eldholm, 1994). They are generally referred to here as hot-spots even though there is disagreement about whether their generation always requires higher than normal temperatures. Typical hot-spot provinces include the northwest Tertiary igneous province of Britain, itself a small part of the magmatism associated with the Iceland hot-spot, and the Deccan Traps in India.

In the oceans, hot-spots are recognised as areas that are shallower than 'normal' ocean floor of the same age. Sometimes they are contemporaneous with spreading, as in Iceland. Other hot-spots postdate the ocean floor by tens of millions of years, as in the Hawaiian–Emperor Seamount Chain, or give rise to oceanic plateaux, such as the Ontong Java plateau in the southwest Pacific.

Normal ocean floor is created by the partial melting of the asthenosphere at oceanic ridges that results in 'standard' ocean floor with a thickness of 7 km (McKenzie and Bickle 1988). The high temperatures at an active ridge reduce the density of the underlying rocks compared with areas some distance from the ridge, causing them to rise above adjacent older ocean floor. Active ridges typically rise to ~2.6 km and old ocean floor becomes deeper as it cools down and isostatic balance is maintained. The relationship between age and depth is well established (Parsons and Sclater, 1977; Stein and Stein, 1992; Hillier and Watts, 2005; Crosby et al., 2006; Zhong et al., 2007). Thus, in general, anything that is significantly less deep than the value given by the standard age–depth relationship must have a thicker than normal crust and, apart from buried continental slivers, such as parts of Kerguelen (Nicolaysen et al., 2001), or areas covered by thick sediment, is interpreted as abnormally thick oceanic crust. The abnormal thickness is attributed to the generation of more basalt than is generated by the adiabatic decompression of normal asthenosphere (the asthenosphere must have been hotter than normal or richer in basaltic components) and hence be a sign of hot-spot activity. The easiest way to identify most oceanic hot-spots is to plot the 2,000-m contour: anything shallower than 2,000 m in the oceans that is not obviously part of a continent is a probable candidate for a hot-spot.

3.4.2 OCEANIC HOT-SPOTS AS AN 'ABSOLUTE' REFERENCE FRAME

Hot-spot activity in the oceans commonly gives rise to linear features. Two of the best known are the Hawaiian–Emperor chain in the Pacific, with its enigmatic bend, and the Ninety East Ridge running south from India (Figure 3.12).

Because hot-spots move more slowly relative to one another than do plates, it has been suggested that hot-spots might provide a so-called 'absolute' reference frame for studying plate motions (e.g., Duncan 1981). A very simple model imagines hot-spots as the surface expression of jets of hot material rising upward from the core–mantle boundary much faster than plates move at the surface or faster than convection in the mantle beneath (e.g., Richards et al., 1989). Hot-spots would therefore provide a reference frame (Figure 3.12) that could in principle eliminate the longitude uncertainty of palaeomagnetism (e.g., Müller et al., 1993). It is unclear what the mean hot-spots frame represents as one goes back in time, nor which hot-spots frame to use: a Pacific hot-spots reference frame differs significantly from one based on Atlantic–Indian Ocean hot-spots (Norton, 2000; Figure 3.13). By contrast, the palaeomagnetic frame

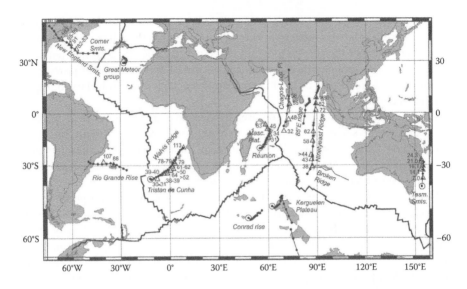

FIGURE 3.12 The Mercator map shows the main hot-spots in the Atlantic and Indian Oceans, modified from Müller et al. (1993). Active hot-spots are open circles. The modelled tracks are shown at 5-Ma intervals as black dots connected by lines. Measured dates are shown as numbers next to triangles by the tracks. There is excellent agreement between the observed and modelled dates and tracks. For details see Müller et al. (1993). (Modified from Müller, R.D., Royer, J.Y. and Lawver, L.A. (1993). *Geology*, 21: 275—278, figure 1.)

provides an estimate of the position of the Earth's spin axis in past time, which is fundamental to palaeogeography. Such a frame differs significantly from that provided by either of the hot-spots frames (Figure 3.14).

Because continental basaltic volcanism rarely leaves a trail of the kind found in the oceans, and oceanic hot-spot trails in the Atlantic and Indian Oceans are poorly known prior to this date, there is only fragmentary knowledge of their spatial distribution prior to 130 Ma (Figure 3.12), although the temporal distribution of flood basalts suggests that hot-spots have been a long-term feature of the Earth's surface (Prokoph et al., 2004, Ernst and Buchan, 1997).

3.5   MAPS AND COMPOSITES

3.5.1   MESOZOIC AND CENOZOIC MAPS

Mesozoic and Cenozoic maps are simply figures showing reassemblies of the continents that have been turned into maps using palaeomagnetic data. These 'maps' are still only rudimentary palaeogeographic maps: they do not show former coastlines, former continental topography, or former bathymetry; rather they show where the continents and oceans were in the past.

In the Mesozoic and Cenozoic the magnetic anomalies of the Atlantic and Indian Ocean ocean floor are the prime source of how the continents should be reassembled. As one goes back to the Triassic, the Pacific Ocean Basin increases in size, but globally the area with known magnetic anomalies on it becomes smaller. Anything

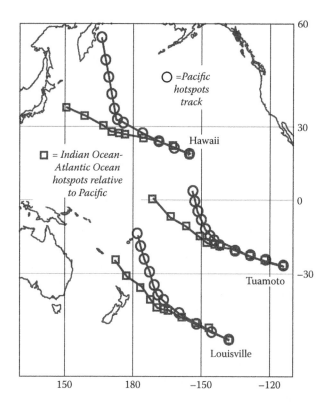

FIGURE 3.13 Mismatch between Atlantic and Pacific hot-spots frames. Three Pacific hot-spots trails—Hawaii, Tuamoto, and Louisville—are shown from 90–0 Ma at 10-Ma intervals, together with the computed positions of the same hot-spots in an Atlantic–Indian hot-spots frame. The trails diverge as one goes back in time, showing that the two frames are incompatible. (Modified from Norton, I.O. 2000, in M.A. Richards, Gordon, R.C. and Van der Hilst, R.D. (eds.), The history and dynamics of global plate motions. *Geophysical Monograph.* Washington, D.C., American Geophysical Union, 121, pp. 339—357, figure 8.)

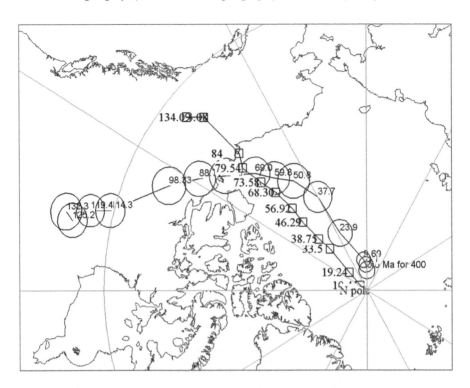

FIGURE 3.14 Hot-spots and palaeomagnetic frame compared. The position of the north pole relative to Africa is shown for the Atlantic–Indian hot-spots from 120–0 Ma and for the mean palaeomagnetic pole from 130–0 Ma. The hot-spots and mean palaeomagnetic poles are displayed at irregular intervals with the corresponding α_{95} circle around each palaeomagnetic pole. Errors are not known for the hot-spots frame, but the two curves are never more than ~5° apart for 90–0 Ma, but rapidly increase to >10° thereafter.

younger than the map did not exist at the time and subduction has removed much that did exist. Some of the subducted ocean floor can be resurrected by assuming spreading was symmetric on the East Pacific Rise. For convenience, the Mesozoic and Cenozoic maps are referred to as palaeocontinental maps. More recent data for making maps are given in Torsvik et al. (2007).

3.5.2 PALAEOZOIC 'MAPS' (OR 'COMPOSITES')

There is no undeformed Palaeozoic ocean floor. Therefore each Palaeozoic continent has to be oriented into its correct palaeolatitude by using its own mean palaeomagnetic pole. Because palaeomagnetism does not give longitude, the separations of the Palaeozoic continents from one another is based on geological or palaeontological criteria. In the Palaeozoic the former oceans are simply the spaces between the continents with little direct evidence to support their former existence. Old orogenic belts, subduction zones, and ophiolites are the main features produced during oceanic closure but the former size of a vanished ocean is difficult to estimate.

The re-evaluation of the palaeogeographical significance of a fossil group can bring about a change in the preferred separation between two or more Palaeozoic continents and hence a change in the shape and position of former Palaeozoic oceans. Of course, each group of workers is likely to have a different opinion on these separations. Palaeozoic palaeocontinental reassemblies are essentially collages of individual Palaeozoic continents projected onto the same map frame and are referred to here as *composites*, rather than maps (e.g., Smith et al., 1973). However, whether Cenozoic, Mesozoic, or Palaeozoic in age, all the maps and composites are simply foundations to which the geographical features need to be added.

3.5.3 Uncertainties in Cenozoic and Mesozoic Maps

As noted above, the global mean α_{95} is less than 2° for Mesozoic and Cenozoic time (Figure 3.11). All the major continents are linked by detailed ocean floor data whose positions are determined by GPS methods and therefore have small positional errors.

There are also uncertainties associated with the extension that has taken place on continental margins during the breakup of Pangaea and its components (see below for details), or the changes that have taken place at active continental margins. The extension on passive continental margins is probably about 100 km for a margin that is 200-km wide such as parts of the eastern North American margin (e.g., Sawyer, 1985). In other areas, the extension may be much higher: Powell et al., (1988) estimated a total of 360-km extension across the Australia–Antarctica margin before the formation of ocean floor (180 km) equivalent to 1.6° of latitude for each margin, assuming symmetric extension. Even this amount of extension is relatively small on a global scale and would be equivalent to drawing the boundaries of the continents on a page-size global map with a medium felt tip pen rather than a sharp pencil.

The main uncertainties in Mesozoic and Cenozoic maps are in areas affected by orogeny. 'Greater India' and 'The Great Alaskan Terrane Wreck' illustrate some of these uncertainties (Figures 3.15 and 3.16).

During the continent–continent collision of India with Eurasia the crust may have been doubled in thickness. A plausible cartoon of how this happened is shown in Figure 3.15. Six hundred km, and perhaps as much as 800 km, have been lost at the surface and now lie under the Himalayas and Tibetan plateau. As one goes back in time, the subducted Indian crust should be pulled out from under Eurasia to form an additional continental fragment. This fragment is sometimes referred to as 'Greater India', although this term has also been used for the present-day extent of the Indian plate, which continues well beyond India itself.

Similar continent–continent collisional processes have probably taken place in all collisional zones (i.e., in most orogenic belts), and therefore, strictly speaking, as one goes back in time, one should pull out the equivalent of a Greater India in every collisional orogenic belt, and add such areas to the appropriate continent to show its shape prior to collision. No global reconstructions routinely make these adjustments, which will generally be much less than required for the Himalayas and Tibet, but might significantly change the outlines of some areas.

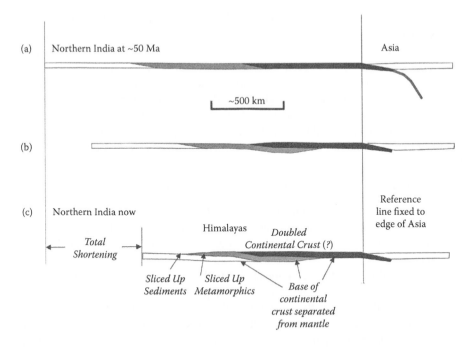

FIGURE 3.15 Himalayas. Three cartoons are shown for a cross-section between Asia and India: (a) at ~50 Ma, (b) at an intermediate age, and (c) the present-day, based on sections in Shackleton et al. (1988). The overall process is simple: crustal shortening of as much as 600 km has taken place, but the results are complex in detail. When India is extracted from Asia this amount of continental crust should be added to the northern edge of India, which is significant on a global scale. Similar corrections are needed for all orogenic belts exposing high-grade metamorphic rocks in which the crust was buried to >30 km, but the magnitude of crust to be extracted is probably generally less than in the case of the Himalayas.

Changes in shape induced by large-scale strike-slip (transform) faulting have been proposed for the western margin of North America during the interval 80–45 Ma (Johnston, 2001). Parts of a large complex collage of island arcs and continental slivers may have been transported some 3,000 km along the margin and bent and folded in a process that has significantly changed the shape of the margin forming what is now the 'Great Alaskan Terrane Wreck' (Figure 3.16).

The motions have juxtaposed faunas and floras that were originally more than 25° of latitude apart that may represent different faunal and floral provinces. Had the displacements not been recognised, invalid palaeobiological interpretations might have been drawn from the juxtaposition of these provinces. Conversely, in other areas, the juxtaposition of disparate faunas and floras may lead to the detection of large-scale movements that had not previously been recognised.

3.5.4   Uncertainties in Palaeozoic Composites

The existence of separate continents in the Palaeozoic that cannot be linked together by ocean floor data amplifies the uncertainties already present in Mesozoic and

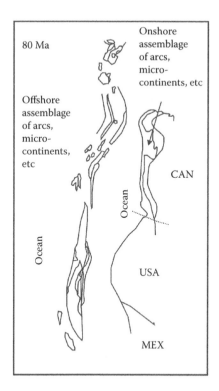 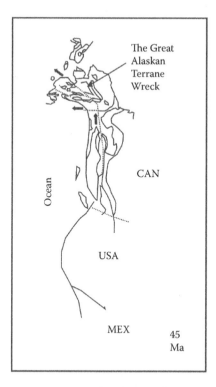

FIGURE 3.16   Great Alaskan Terrane Wreck. Two cartoons are shown of the western North American continental margin at 80 Ma and 45 Ma. During this period complex motions take place along the margin that may have resulted in lateral transport of some fragments by >2500 km. (Modified from Johnston, S.T. (2001). *Earth and Planetary Science Letters*, 193: 259-272. With permission from Elsevier.)

Cenozoic maps. Because each Palaeozoic continent is an entity, the amount of palaeomagnetic data available is much less than that used for the global mean pole determinations of the Mesozoic and Cenozoic, where data from all the stable continents are amalgamated into a single dataset (e.g., Schettino and Scotese, 2005). The problems are illustrated by Cenozoic and Mesozoic Siberia, where α_{95} may be several times higher than that for the global mean pole (Figure 3.11).

If two continents are repositioned by palaeomagnetism, rather than being part of a single global reassembly, then the α_{95}s of both continents have to be taken into account, again increasing the uncertainties. Thus some parts of a given Palaeozoic composite are relatively well determined and others may be poorly determined.

3.5.5   Palaeozoic Composites and Palaeomagnetism: Pangaea A and B

The classic Pangaea of early Jurassic time, so-called Pangaea-A, is the least-squares fit of the 1,000-m submarine contour (Bullard et al., 1965), which is very similar to the fit of the 2,000-m contour (Figure 3.17A). It is supported by the fit of the oldest magnetic anomalies (Klitgord and Schouten, 1986) and by the late Triassic/

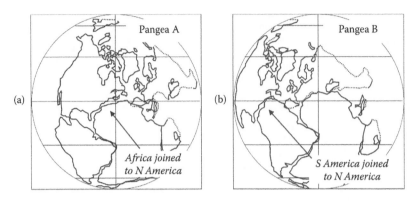

FIGURE 3.17 Pangaea A and B. (a) Pangaea A is the classic Pangaea in which Africa tucks into eastern North America. It is well supported by the earliest ocean-floor data, best fits of the continental margin, and late Triassic to early Jurassic palaeomagnetic data. (b) Pangaea B attempts to bring the mean Permian palaeomagnetic poles of Laurussia and Gondwana into coincidence (Morel and Irving, 1981). This can be done only by moving Gondwana northeast relative to its Pangaea A position. An alternative view is to keep the Pangaea A position and reinterpret the palaeomagnetic data as representing a strongly nondipole field (e.g., Torsvik and Van der Voo, 2002). A third view is that the Permian palaeomagnetic data have been remagnetised.

early Jurassic palaeomagnetic data from Laurussia and Gondwana (Livermore et al., 1986). Laurussia and Gondwana collided during the early Mississippian to late Permian (~360 Ma to ~260 Ma) to form Pangaea (Hatcher et al., 1989, Figure 2). Early Permian folds are the youngest structures formed by the collision. One would therefore expect that most of the Permian palaeomagnetic data from Laurussia and Gondwana would also fit well on Pangaea-A. They do not (Morel and Irving, 1981): modern data and reconstructions show that there is a mismatch of 7° in the mean poles at 250 Ma, more than can be accounted for by the α_{95}s (Torsvik and Van der Voo, 2002). How to interpret this mismatch is a long-standing problem for which several solutions have been proposed that are not discussed in detail here, although the most important implications are outlined.

A common solution is to suppose that the fundamental assumption of palaeomagnetism applies (the GADF model) and that during the period of mismatch Gondwana lay northeast of its Pangaea-A position to form Pangaea-B (Morel and Irving, 1981) where the mean poles can be brought into coincidence and South America lies opposite eastern North America (Figure 3.17B). However, this match of the palaeomagnetic data can be attained only by largely strike-slip motion of several thousand kilometres (e.g., Torsvik and Van der Voo, 2002) along a major transform (or transforms) that has never been found (e.g., Smith and Livermore, 1991).

A second alternative assumes that the GADF model breaks down and that Pangaea-B never existed (Torsvik and Van der Voo, 2002). If this is true, then the foundation for making Palaeozoic composites from palaeomagnetic data is unsound, and introduces additional latitudinal uncertainties, that could be at least 7°, into the position of individual pre-Mesozoic continents, whatever their age.

A third alternative, which in the writer's view has not been thoroughly investigated, is that the discrepant Permian data have been remagnetised. Why there is mismatch between the Permian palaeomagnetic data and Pangaea-A is a major unsolved problem in global tectonics and palaeomagnetism.

3.5.6 FUSULINID DISTRIBUTION: EVIDENCE FOR TRANS-PACIFIC MIGRATION OF ISLAND ARCS?

The problems of repositioning Palaeozoic island arcs and continental fragments are illustrated by two examples: the first from the distribution of the fusuline-bearing rocks of western North America; the second, the reconstruction of the complex collage that now makes up the Altaids of central Asia.

The first case has given rise to suggestions that some Permian rocks have been transported several thousand kilometres relative to their present positions on the North American margin. Palaeomagnetic data support this interpretation, but because of the longitudinal uncertainty of palaeomagnetic data, a range of models is possible, with two end members. One end member proposes that the fusulinds were moved by transform faulting along the margin (e.g., Cowan et al., 1997), as in the late Mesozoic-early Cenozoic Alaskan Terrane Wreck, and have never been far from that margin (Figure 3.18).

FIGURE 3.18 Speculative late Palaeozoic position of some terranes carrying Fusulinacean faunas. The terranes (Wrangellia, Alexander, Cache Creek, Stikinia, and Quesnellia) now form part of the western North American margin (Figure 3.16). The terranes would have migrated across much of the late Palaeozoic Pacific basin. (Adapted from Gabrielse, H., Monger, J.W.H., Wheeler, J.O., and Yorath, C.J. (1991), in H. Gabrielse, and C.J. Yorath (eds.) *Geology of the Cordilleran Orogen in Canada.* figure 2.10.)

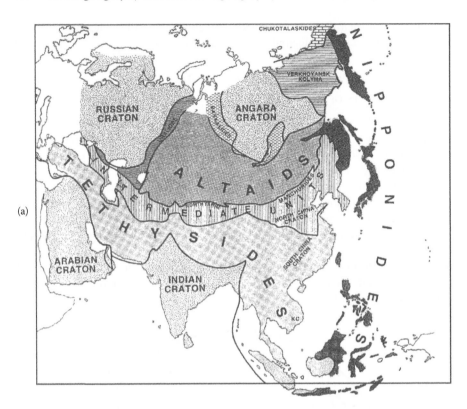

FIGURE 3.19   (a) The Altaids are a Palaeozoic collage of island arcs, ophiolites, and continental fragments bordering the southern Russian (= Baltica) and Siberian (= Angara) cratons (= shields). (From Sengör, A.M.C., and Natal'in, B.A. (1996) in A. Yin and M. Harrison (eds.) *The tectonic evolution of Asia: Cambridge, England.* Reprinted with the permission of Cambridge University Press, pp. 486—640, figure 21.7).   *Continued*

The second end member postulates that during the Early Carboniferous what are now fragments of fusulinid-bearing limestones in Mesozoic island arcs (or microcontinents) in western North America may have formed oceanic islands and shallow-water oceanic areas that lay quite close to Australia and Antarctica. They then migrated several thousand kilometres east by Late Permian time (Carter et al., in Gabrielse and Yorath 1991, figure 2.10), or alternatively may have migrated from the eastern Tethys across the Pacific basin. The principle of least astonishment suggests the transform origin is more likely, but this is a philosophical rather than a scientific principle.

3.5.7   THE ALTAID COLLAGE

The Altaids of central Asia are a complex of the late Precambrian and earlier Palaeozoic island arc and continental fragments that was assembled into a collage by Carboniferous time (Figure 3.19a). One end member of a range of possible Altaid models, and the most detailed presented to date (Sengor and Natal'in, 1996), suggests

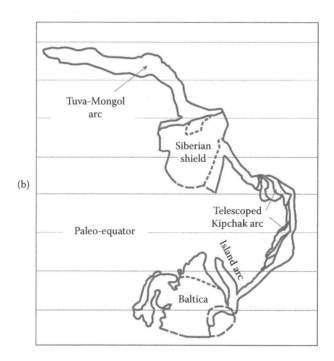

FIGURE 3.19 (*Continued*)   (b) Speculative late Ordovician reconstruction showing two major arcs: the Tuva–Mongol arc forming an elongate finger pointing west (in palaeoco-ordinates) from the Siberian shield; and the Kipchak arc, connecting the Siberian shield to Baltica. (From Sengör, A.M.C., and Natal'in, B.A. (1996), figure 21.32. Reprinted with the permission of Cambridge University Press.)   *Continued*

that the Altaids evolved from two elongate island arcs, the Kipchak arc, joining the East European craton to the Siberian craton, and the Tuva–Mongol arc that extended as a long curved finger pointing westward (in palaeoco-ordinates) from the northern Siberian craton (Figure 3.19b). The postulated stages in the assembly are complicated and only two intermediate steps are illustrated here (Figures 3.19a–c). Quite different models exist to account for the Altaids, but space prohibits their presentation here. The essential point is that many features of Palaeozoic orogenic belts are poorly understood, are not well constrained, and may involve relative motions between their components and between the adjacent stable continents of thousands of kilometres, giving rise to significant additional uncertainties in Palaeozoic composites.

3.6   CONTINENTAL MOTIONS AND BIOLOGICAL EVOLUTION

The errors and uncertainties in global reconstructions have been outlined above. In the last section of this chapter some of the tectonic effects that may be important for biological evolution are discussed in more detail. Hot-spots cause some of the most interesting effects; they appear to be important for continental breakup; for the open-ing of oceanic gateways; for indirectly causing rapid, short-lived global temperature changes; and for their intrinsic damage to the environment by their extraordinary

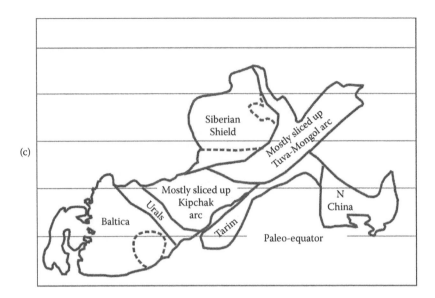

(c)

FIGURE 3.19 (*Continued*) (c) Speculative late Carboniferous reconstruction of the Altaids. The arcs have collapsed, wrapping themselves around Baltica and Siberia. (From Sengör, A.M.C., and Natal'in, B.A. (1996), figure 21.37. Reprinted with the permission of Cambridge University Press.)

volcanic activity. The migration of island arcs may also bring about significant biological changes.

3.6.1 HOT-SPOTS AND CONTINENTAL BREAKUP

Most Mesozoic and Cenozoic passive continental margins show at least one example of intense basaltic activity along at least one part of every margin just prior to breakup (White, 1992; Menzies et al., 2002), although the conjugate margins of Australia and Antarctica formed in the absence of any major igneous activity. After breakup these become passive volcanic continental margins that may pass laterally into passive nonvolcanic margins.

Most of this igneous activity is attributed to material rising up from sources as deep as the core–mantle boundary (mantle plumes) or as shallow as the lower lithosphere (White and McKenzie, 1995; Courtillot et al., 2003). The initial effect of a plume on undeformed lithosphere is to heat it, raising its surface but reducing its thickness and weakening it (White and McKenzie, 1989). Uplift acts like oceanic ridge-push. If the lithosphere is already under extensional stress (e.g., from slab-pull by a subduction zone on the edge of a plate), the increased stresses resulting from uplift, together with the weakening caused by lithospheric thinning, previous extension, and membrane stresses, may be sufficient to break it apart (Hill, 1991; Bott, 1992; White, 1992). Hot-spots are therefore likely to be important in defining the location and timing of the breakup of a continent.

3.6.2 HOT-SPOTS AND OCEANIC GATEWAYS

When continents separate, they open oceanic gateways, with the possible evolutionary consequences outlined above. In addition, the uplift may raise nearby continental shelves, decomposing the methane ices (clathrates) present on the shelves. This mechanism is suspected to be a prime cause for the remarkable temperature excursion near the Palaeocene/Eocene boundary (Maclennan and Jones, 2006). During an interval estimated to be ~20 kyr the global oceanic temperature rose by an estimated 3–10°C, which must have had a biological impact. Of course, this process can in theory take place independently of hot-spots and oceanic gateways.

3.6.3 FLOOD BASALTS AND MASS EXTINCTIONS

Flood basalts may also cause mass extinctions. Significant biotic changes take place at the times of several flood basalt eruptions or during the formation of large igneous (basaltic) provinces (Wignall, 2001). There are also direct environmental effects including climatic cooling from sulphuric acid aerosols, warming from CO_2 and SO_2 gases, and resulting acid rain. In the oceans, the formation of large submarine plateaus such as the Ontong Java Plateau may change ocean chemistry, circulation, and oxygen content.

3.6.4 LONG-TERM EVOLUTION OF ISLAND ARCS

Many island arcs may have started their existence at an ocean–continent convergence zone as part of a continental arc. As the hinge line of the subduction zone retreats, the space created is filled by the upward rise and partial melting of the underlying asthenosphere. The lithosphere under the arc is hot and therefore weak, and may break off as a continental sliver at the arc, which then migrates and becomes an island arc with a back-arc basin behind it.

As time passes, the island arc may itself be split, leaving behind a *remnant arc*, with the actively spreading back-arc basin immediately behind the active arc. The arc may continue to migrate oceanward. For example, starting next to Australia, successive breakup and migration may have created the present complex pattern of fragments now consisting of the Lord Howe Rise (oldest formed fragment), New Caledonia, the Lau–Colville Ridge, and the Tonga–Kermadec Islands (Figure 3.20). During their migration, island arcs may change their shape, apparently bending with time (e.g., the Marianas arc off western Asia). Unravelling the tectonics of such regions is not easy (e.g., Hall, 1997).

What is unclear is whether such arcs cross major oceans, as suggested above for the Permian fusulinid distributions in parts of the Pacific Basin, whether they become entangled with one another, as in the case of the Altaid collage, or whether they simply collapse back onto the continental margin from which they originally were spawned.

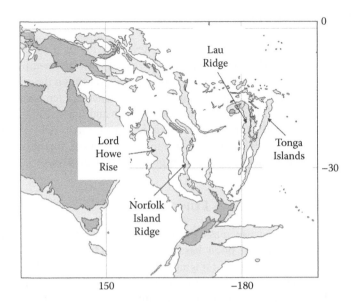

FIGURE 3.20   Present-day coastlines and 2,000-m submarine bathymetry of the southwest Pacific. As one goes back in time, the Tonga Islands join onto the Lau Ridge; the Tonga–Lau fragment then joins onto the Norfolk Island Ridge; this composite fragment then joins onto the Lord Howe Rise and all of them then join Australia. Of course, much of the igneous activity in the younger arcs is absent in the older composite fragments. Other fragments (not labelled) have a similar history. This is a possible present-day analogue in the early stages of those models that see fragments of the North American Pacific margin originating in the Pacific Basin or even Australasia (Figure 3.18).

3.7   PALAEOBATHYMETRY AND PALAEOTOPOGRAPHY

3.7.1   Ocean Floor Bathymetry

The ultimate aim of palaeogeography is to reconstruct the geography of the past in as much detail as possible. It will never be possible to make maps with the detail of the present day, but major steps can now be made, particularly in reconstructing ocean floor bathymetry. The depth of ocean floor depends on its age, with well-known expressions relating age to depth (see above). These relationships apply only to ocean floor that has not been covered by significant sediment. Where it has, the effects of the sediments on depth can be removed if the thickness, age, and lithology of the sediments are known. The corrections are best applied to gridded rather than vector data. Applying the standard age–depth equations to the corrected bathymetry reproduces the former submarine bathymetry.

3.7.2   Effects of Hot-Spots on Oceanic Bathymetry

Hot-spots on the ocean floor perturb the age–depth relationship by reducing the depth of the ocean floor and a second correction must be applied. How to do this is unclear, principally because there is as yet no general theory for the growth of a hot-spot and how that growth will be reflected in the depth of the ocean floor. For example, as the hot-spot grows, the ocean floor becomes shallower, but as the hot-spot source moves elsewhere the older part of the hot-spot trail cools and subsides, as in the Emperor seamounts on the older part of the Hawaiian–Emperor chain in the Pacific. The correction will be important when a hot-spot trail starting at the edges of conjugate continents causes significant changes in oceanic circulation, or has subaerial volcanoes along its length that provide 'land bridges' for migration.

3.7.3 EFFECTS OF CHANGES IN THE MEAN AGE OF THE OCEAN FLOOR ON SEA LEVEL

The depth of water above all spreading ridges that are distant from any hot-spots is about 2.6 km. The young ocean floor cools, subsides, and in the absence of subduction, the best-fitting age–depth relationship suggests it would eventually reach a depth of about 5.6 km.

Were the average ocean floor age younger than it is today, then the depth of water above a newly formed ridge would be greater. Global sea level would have been higher and the land area would have been smaller. The converse is also true. These long-term variations clearly should have important biological effects. For the past 180 Ma there appears to have been an initial slower phase of sea-level rise (by 90 to 110 m), reaching a maximum between 120 and 50 Ma, followed by a faster phase of sea-level drop (Cogne et al., 2006). When the effects of oceanic plateau formation and ice cap development are added, the predicted sea-level curve fits remarkably well with the first-order variations of sea-level change inferred from stratigraphic and geophysical data. Thus, the changes in mean age of the oceanic lithosphere (varying between 56 and 62 ± 0.2 Myr), account for ~70% of first-order changes in sea-level.

3.7.4 PALAEOCOASTLINES

The shoreline is the palaeogeographical feature that is commonly highlighted on global palaeogeographical maps. The coastline is an ephemeral geographical feature: it may migrate backwards and forwards across large areas of a continent with geological rapidity. However, it provides a very sensitive indicator of changes in base level, particularly in areas of low gradient where a small relative rise or fall in sea level will cause a correspondingly large movement of the coastline which in turn will have some palaeobiological significance.

Ideally, placing a palaeocoastline on a map is no more than drawing the boundary between the extent of marine sedimentation on one side of a line and continental sedimentation on the other, but adequate data are available mostly in western Europe and North America. In other areas, the low density of the available data means that palaeocoastlines at the global scale must be regarded as very speculative (Smith et al., 1994). Even where relatively detailed data are available, it is very difficult to infer the former coastlines of deformed microcontinents that may also have been partially subducted.

Smith et al. (op. cit. Figure 8) calculated the land area on their Mesozoic and Cenozoic maps. With few exceptions, the gross fluctuations in land area are quite similar to the changes in sea-level inferred by Haq et al. (1988) from seismic reflection profiles and stratigraphic data, at least on a time-scale of about 25 Ma or more. Although such a relationship is to be expected because a rise in sea level leads to a reduction in land area, the sea-level curve and that of Haq et al. have been derived from quite different datasets. Haq et al. (1988) suggest changes may exceed 200 m in the late Cretaceous. Cogne et al. (2006), using different and more recent data and more recent global reconstructions, envisage a maximum of 170–180 m, although other estimates of the maximum rise are lower (e.g., Watts and Thorne, 1984).

3.7.5   CONTINENTAL TOPOGRAPHY

The variations in continental topography are more complex and have several causes, few of which are well understood. In most cases, the best that can be done with present-day understanding is to show past topography as ill-defined high areas and display their variations in time in a geologically plausible manner.

3.7.6   MODELLING PAST CLIMATES

Palaeoclimate models show the likely climate in past time (e.g., Barron and Moore, 1994). The computer models themselves are constantly being improved, particularly in recent years, by coupling the oceanic and atmospheric circulations. They require a great deal of computing power to run: only relatively recently has it been possible to model the major eddies in the surface ocean waters (e.g., the Gulf Stream) in a global model (Semtner and Chervin, 1992); the grid spacing was previously too coarse.

Global circulation models require a knowledge of atmospheric composition, particularly CO_2, a topic beyond the scope of this chapter. Those parts of the model results that depend on the global reconstructions are no better than the data on which they are based. Thus, the global reconstruction input to Cenozoic and Mesozoic models is reasonably well constrained, except where past continental topography may have been important, or times when oceanic gateways were opening or closing. If the output from the model seems to be at variance with the geological data, the errors are less likely to be in the global reconstructions than in some other data. The input to Palaeozoic models has much larger uncertainties and Palaeozoic global climate models are probably still in their infancy.

SUMMARY

Apart from fragments caught up in orogenic belts, the past positions of continents in the Cenozoic and Mesozoic are relatively well known. Uncertainties are much larger for the Palaeozoic continents, partly because of the longitudinal ambiguity of palaeomagnetic data, partly because of the palaeomagnetic data itself, and partly because the nature of the Palaeozoic magnetic field is unclear. The palaeobathymetry for the Cenozoic and, to a lesser extent, Mesozoic oceans can be estimated, although there are significant perturbations due to hot-spots. Palaeozoic palaeobathymetry is highly uncertain. Palaeotopography is uncertain for the entire Phanerozoic. Hot-spots are important for biological evolution: they are probably a critical factor in breaking up continents, thereby isolating faunas and floras. They also create short-lived ocean gateways. The associated uplift of nearby continental shelves, together with changes in the ocean currents, may cause the breakdown of methane ices, which can give rise to transient spikes in global temperatures that may influence evolution. In addition, significant biotic changes, including some mass extinctions, appear to be contemporaneous with flood basalt eruptions.

ACKNOWLEDGEMENTS

The author thanks the organisers of the *Conference on Palaeogeography and Palaeobiogeography: Biodiversity in Space and Time,* for the invitation to present this chapter. He also thanks Ronald Blakey and an anonymous reviewer for their helpful comments, and Lawrence Rush, of Cambridge Paleomap Services Limited, who wrote the ATLAS program to display many of the maps.

REFERENCES CITED

Barron, E.J. and Moore, G.T. (1994) *Climate modelling in paleoenvironmental analysis. SEPM short course,* no. 33. Society for Sedimentary Geology, Tulsa, OK.

Bird, P. (2003) An updated digital model of plate boundaries. *Geochemistry Geophysics Geosystems, 4,* 127, doi:10.1029/2001GC000252.

Bott, M.H.P. (1992) The stress regime associated with continental break-up. *Special Publication of the Geological Society of London,* 68: 125–136.

Bullard, E., Everett, J.E., and Smith, A.G. (1965) The fit of the continents around the Atlantic. *Philosophical Transactions of the Royal Society of London, A258*: 41–51.

Carter, E.S., Orchard, M.J., Ross, C.A., Ross, J.R.P., Smith, P.L., and Tipper, H. (1991) Part B, Paleontological Signatures of Terranes. In *Geology of the Cordilleran orogen of Canada, Geology of Canada,* no. 4 (eds. H. Gabrielse and C.J. Yorath), Geological Survey of Canada, Ottawa, pp. 28–38.

Chian, D., Keen, C.E., Reid, I., and Louden, K.E. (1995) Evolution of non-volcanic margins: New evidence from the conjugate margins of the Labrador Sea. *Geology,* 21: 589–592.

Coffin, M.F. and Eldholm, O. (1994) Large igneous provinces: Crustal structure, dimensions, and external consequences. *Reviews of Geophysics,* 32, 1–36.

Cogne, J.P., Humler, E., and Courtillot, V. (2006) Mean age of oceanic lithosphere drives eustatic sea-level change since Pangea breakup. *Earth and Planetary Science Letters,* 245, 115–122.

Courtillot, V., Davaille, A., Besse, J., and Stock, J. (2003) Three distinct types of hotspots in the Earth's mantle. *Earth and Planetary Science Letters,* 205: 295–308.

Cowan, D.S., Brandon, M.T., and Garver, J.I. (1997) Geologic tests of hypotheses for large coastwise displacements—a critique illustrated by the Baja British Columbia controversy. *American Journal of Science,* 297, 117–173.

Crosby, A.G., McKenzie, D., and Sclater, J.G. (2006) The relationship between depth, age, and gravity in the oceans. *Geophysical Journal International,* 166: 553–573.

Duncan, R.A. (1981) Hotspots in the Southern Ocean - An absolute frame of reference for motions of the Gondwana continents. *Tectonophysics,* 74, 29–42.

Ernst, R.E. and Buchan, K.L. (1997) Giant radiating dyke swarms: Their use in identifying pre-Mesozoic large igneous provinces and mantle plumes. In *Large Igneous Provinces* (eds. J.J. Mahoney and M.F. Coffin), American Geophysical Union, Washington DC.

Fowler, C.M.R. (2005) *The Solid Earth: An Introduction to Global Geophysics*, Cambridge University Press, Cambridge, UK.

Gabrielse, H., Monger, J.W.H., Wheeler, J.O., and Yorath, C.J. (1991) Morphogeological belts, tectonic assemblages, and terranes. In *Geology of the Cordilleran Orogen in Canada* (eds. H. Gabrielse, and C.J. Yorath). Ottawa, Ontario: Minister of Supply and Services Canada.

Gradstein, F.M., Ogg, J., and Smith, A.G. (2004) *A Geologic Time Scale 2004*, Cambridge University Press, Cambridge, UK.

Hall, R.H. (1997) Reconstructing SE Asia. *Special Publication of the Geological Society of London*, 126, 11–23.

Haq, B.U., Hardenbol, J., and Vail, P.R. (1988) Mesozoic and Cenozoic chronostratigraphy and eustatic cycles. In *Sea-Level Changes: An Integrated Approach. Special Publication 42.* (eds. C.K. Wilgus, B.S. Hastings, H. Posamentier, J. Wagoner, C.A. Ross, and C.G.S.C. Kendall), Society of Economic Paleontologists and Mineralogists, Tulsa, OK, p. 71–108.

Hatcher, R.D., Jr., Thomas, W.A., and Viele, G.W. (1989) *The Appalachian-Ouachita Orogen in the United States: The Geology of North America F2*, The Geological Society of America, Boulder, CO.

Hill, R.I. (1991) Starting plumes and continental break-up. *Earth and Planetary Science Letters,* 104: 398–416.

Hillier, J.K. and Watts, A.B. (2005) Relationship between depth and age in the North Pacific Ocean. *Journal of Geophysical Research,* 110: B02405, 22, doi:10.1029/2004 JB003406.

Johnston, S.T. (2001) The Great Alaskan Terrane Wreck: Reconciliation of paleomagnetic and geological data in the northern Cordillera. *Earth and Planetary Science Letters,* 193: 259–272.

Jupp, P.E. and Kent, J.T. 1987. Fitting smooth paths to spherical data. *Applied Statistics,* 36: 34–36.

Klitgord, K.D. and Schouten, H. (1986) Plate kinematics of the central Atlantic. In *The Western North Atlantic Region, Volume M: Geology of North America* (eds. P.R. Vogt and B.E. Tucholke), Geological Society of America, Boulder, CO, pp. 351–378.

Livermore, R.A., Smith, A.G., and Vine, F.J. 1986. Late Palaeozoic to early Mesozoic evolution of Pangea. *Nature,* 322:162–165.

Maclennan, J. and Jones, S.M. (2006) Regional uplift, gas hydrate dissocation, and the origins of the Paleocene-Eocene Thermal Maximum. *Earth and Planetary Science Letters*, 245, 65–80

McElhinny, M.W. and McFadden, P.L. (2000) *Palaeomagnetism: Continents and Oceans.* Academic Press, San Diego.

McKenzie, D. and Bickle, M.J. (1988) The volume and composition of melt generated by extension of the lithosphere. *Journal of Petrology,* 29: 625–679.

Menzies, M.A., Klemperer, S.L., Ebinger, C.J., et al. (2002) Characteristics of volcanic rifted margins. In *Volcanic rift margins. Geological Society of America Special Paper* (eds. M.A. Menzies, et al.), Geological Society of America, Boulder, CO, 362, pp. 1—14.

Morel, P. and Irving, E. (1981) Palaeomagnetism and the evolution of Pangaea. *Journal of Geophysical Research*, 86, 1858–1872.

Müller, R.D., Gaina, C., Tikku, A., Mihut,D., Cande, S.C., and Stock, J.M. (2000) Mesozoic/ Cenozoic Tectonic Events Around Australia. *Geophysical Monograph*, 121: 161–188.

Müller, R.D., Royer, J.Y., and Lawver, L.A. (1993) Revised plate motions relative to the hotspots from combined Atlantic and Indian Ocean hotspot tracks. *Geology*, 21: 275–278.

Nicolaysen, K., Bowring, S., Frey, F., Weis, D., Ingle, S., Pringle, M.S., and Coffin, M.F. (2001) Provenance of Proterozoic garnet-biotite gneiss recovered from Elan Bank, Kerguelen Plateau, southern Indian Ocean. *Geology*, 29: 235–238.

Norton, I.O. (2000) Global hotspot reference frames and plate motion. In *The History and Dynamics of Global Plate Motions. Geophysical Monograph* (eds. M.A. Richards, R.C. Gordon, and R.D. Van der Hilst), American Geophysical Union, Washington, DC, 121, pp. 339–357.

Parsons, B. and Sclater, J.G. (1977) An analysis of the variation of ocean floor bathymetry and heat flow with age. *Journal of Geophysical Research*, 82, 803–827.

Pickering, K.T. and Smith, A.G. (2001) Oceanic gateways as a critical factor in initiating pre-Mesozoic glaciations. AGU Fall Meeting, Program and Abstracts, General Paleoceanography and Paleoclimatology Contributions IV Session: San Francisco, American Geophysical Union, PP42B-0523.

Powell, C.M., Roots, S.R., and Veevers, S.J. (1988) Pre-breakup continental extension in East Gondwanaland and the early opening of the eastern Indian Ocean. *Tectonophysics*, 155, 261–283.

Prokoph, A., Ernst, R.E., and Buchan, K.L. (2004) Time-series analysis of large igneous provinces; 3500 Ma to present. *Journal of Geology*, 112: 1–22.

Richards, M.A., Duncan, R.A., and Courtillot, E. (1989) Flood basalts and hot-spot tracks: Plume heads and tails. *Science*, 246, 103–107.

Robb, M.S., Taylor, B., and Goodliffe, A.M. (2005) Re-examination of the magnetic lineations of the Gascoyne and Cuvier Abyssal Plains, off NW Australia. *Geophysical Journal International*, 163: 52–55.

Rusby, R.I. and Searle, R.C. (1995) A history of the Easter microplate, 5.25 Ma to present. *Journal of Geophysical Research*, 100: 12617–12640.

Sawyer, D.S. (1985) Total tectonic subsidence: A parameter for distinguishing crust type at the U.S. continental margin. *Journal of Geophysical Research*, 90: 7751–7769.

Scheck-Wenderroth, M., Raum, T., Faleide, J.I., Mjelde, R., and Horsfield, B. (2007) The transition from the continent to the ocean: A deeper view on the Norwegian margin. *Journal of the Geological Society, London*, 164: 855–868.

Schettino, A. and Scotese, C.R. (2005) Apparent polar wander paths for the major continents (200 Ma to the present day); A palaeomagnetic reference frame for global plate tectonic reconstructions. *Geophysical Journal International*, 163: 727–759.

Semtner, A.J. and Chervin, R.M. (1992) Ocean general-circulation model from a global eddy-resolving model. *Journal of Geophysical Research*, 97: 5493–5550.

Sengör, A.M.C. and Natal'in, B.A. (1996) Paleotectonics of Asia: Fragments of a synthesis. In *The tectonic evolution of Asia* (eds. A. Yin and M. Harrison), Cambridge University Press, Cambridge, UK, pp. 486–640.

Shackleton, R.M. (ed.) (1988) *Tectonic evolution of the Himalayas and Tibet. Philosophical Transactions of the Royal Society*, 326: London, The Royal Society.

Smith, A.G. and Livermore, R.A. (1991) Pangea in Permian to Jurassic time. *Tectonophysics*, 187, 135–179.

Smith, A.G. and Pickering, K.T. (2003) Oceanic gateways as a critical factor to initiate icehouse Earth. *Geological Society of London*, 160: 337–340.

Smith, A.G., Briden, J.C., and Drewry, G.E. (1973) Phanerozoic world maps. In *Organisms and Continents Through Time* (ed. N.F. Hughes), Palaeontological Association, London, 12, 142.

Smith, A.G., Smith, D.G., and Funnell, B.M. (1994) *Atlas of Mesozoic and Cenozoic Coastlines*, Cambridge University Press, Cambridge, UK.

Srivastava, S.P. and Roest, W.R. (1999) Extent of oceanic crust in the Labrador Sea. *Marine and Petroleum Geology*, 16: 65–84.

Stein, C.A. and Stein, S. (1992) A model for the global variation in oceanic depth and heat flow with lithospheric age. *Nature*, 359: 123–129.

Torsvik, T. H., Müller, R. D., Van der Voo, R., Steinberger, B., & Gaina, C. (2007). Global plate motion frames: Toward a unified model. *Reviews of Geophysics*, 46, 1–44.

Torsvik, T.H. and Van der Voo, R. (2002) Refining Gondwana and Pangea palaeogeography: Estimates of Phanerozoic non-dipole (octupole) fields. *Geophysical Journal International*, 151: 771–794.

Torsvik, T.H., Smethurst, M.A., Van der Voo, R., Trench, A., Abrahamansen, N., and Halvorsen, E. (1992) Baltica. A synopsis of Vendian-Permian palaeomagnetic data and their palaeo-tectonic implications. *Earth Science Reviews*, 33: 133–152.

Van der Voo, R. (1988) Paleozoic paleogeography of North America, Gondwana, and intervening displaced terranes: Comparisons of paleomagnetism with paleoclimatology and biogeographical patterns. *Geological Society of America Bulletin*, 100: 311–324.

Van der Voo, R. (1990) Phanerozoic paleomagnetic poles from Europe and North America and comparisons with continental reconstructions. *Reviews of Geophysics*, 28:167–206.

Watts, A.B. and Thorne, J. (1984) Tectonics, global changes in sea-level and their relationship to stratigraphical sequences at the US Atlantic continental margin. *Marine and Petroleum Geology*, 1: 319–339.

White, R.S. (1992) Magmatism during and after continental break-up. *Special Publication of the Geological Society of London*, 68: 1–16.

White, R.S. and McKenzie, D.P. (1989) Volcanism at rifts. *Scientific American*, 261: 74–83.

White, R.S. and McKenzie, D. (1995) Mantle plumes and flood basalts. *Journal of Geophysical Research*, 100: 17543–17585.

Wignall, P.B. (2001) Large igneous provinces and mass extinctions. *Earth-Science Review*, 53: 1–33.

Zhong, S., Ritzwoller, M., Shapiro, N., Landuyt, W., Huang, J., and Wessel, P. (2007) Bathymetry of the Pacific plate and its implications for the thermal evolution of lithosphere and mantle dynamics. *Journal of Geophysical Research*, 112: B06412, 18 PP doi:10.1029/2006JB004628.

4 Boundaries and Barriers of North American Warm Deserts

An Evolutionary Perspective

David J. Hafner and Brett R. Riddle

CONTENTS

4.1 INTRODUCTION

Palaeontologists have long appreciated the importance of placing fossils and fossil lineages in their proper geological and palaeoenvironmental context, and as a consequence the disciplines of palaeontology and historical geology have developed in tandem. In contrast, a significant proportion of systematic biologists have historically concentrated on determining the evolutionary relationships amongst living forms without consideration of the changing landscape in which their ancestral lineages evolved. In addition, a longstanding dogma, particularly amongst vertebrate zoologists, was that extant species had evolved during the Pleistocene (e.g., Orr, 1960; Savage, 1960; Findley, 1969; Hubbard, 1974; Morafka, 1974; Schmidly et al., 1993). If an attempt was made to consider the biogeographic history of extant forms and their ancestral lineages, it was usually expressed as an ad hoc addendum to a systematic study, and it was set in a geologically static landscape across which flora and fauna were pushed by the climatic cycles of the Pleistocene glaciations.

The past 30 years have witnessed a major revolution in historical biogeography (Lomolino et al., 2005). The key components of this revolution were: (1) explanation and rapid acceptance of plate tectonics, the mechanism for continental drift (e.g., Hess, 1962); (2) development of rapid and inexpensive methods to directly sequence DNA (e.g., Mullis et al., 1986); and (3) advent of rigorous methods of studying historical biogeography (reviewed in Riddle and Hafner, 2004, 2007). These latter included initial approaches for incorporating phylogenetic information into studies of biogeographic history (e.g., Hennig, 1950, 1966); for inferring past distributions from present ones, and considering patterns of multiple taxa simultaneously (Croizat, 1952, 1958, 1960, 1964); and for reconstructing phylogenies in a cladistic framework (e.g., Platnick and Nelson, 1978). As the laboratory and theoretical methods have advanced, it is increasingly obvious that many extant species are far older than the Pleistocene and that to understand the evolution of lineages and of regional biotas it is necessary to delve far deeper into geological history.

Elsewhere (Riddle and Hafner, 2004, 2007) we have reviewed the specific attributes of phylogeography and particularly comparative phylogeography as an important and progressive subdiscipline within historical biogeography that is particularly useful for reconstructing the biogeographic history of biotas over a Neogene time frame (23 Ma). Phylogeography is a relatively new subdiscipline, introduced by Avise et al. (1987) and defined by Avise (2000) as 'a field of study concerned with the principles and processes governing the geographic distributions of genealogical lineages, especially those within and amongst closely related species'. This new approach used emerging molecular phylogenetic techniques to examine temporally shallow biogeographic histories across gene lineages. By examining phylogeographic patterns of multiple, co-distributed taxa for common historical biogeographic patterns, we can now reconstruct the historical assembly of entire biotas, and assess the relative contributions of vicariance and dispersal to that assembly process. Phylogeography continues to rely on historical geology and palaeontology for critical clues as to the evolutionary response of plants and animals to the dramatic landscape alternations and climatic fluctuations of the Neogene, but offers a complementary predictive tool that reveals previously hidden historical events and directs attention at specific areas most promising for further geological and palaeontological study. We have outlined a specific protocol and rationale that interactively and sequentially employ phylogeographic and historical biogeographic analyses (Riddle and Hafner, 2006a), using as an example our analysis of North American warm deserts. We offer here a more detailed historical and geographic review of these aridlands as a case study illustrating the predictive powers of the comparative phylogeographic approach.

Separately and together, we have been studying the historical development of aridlands biota in western North America for nearly three decades. Our particular focus has been on arid-adapted rodents as model organisms for the evolution of regional deserts. It has been our working thesis that resolution of a shared phylogeographic history within a selected subset of desert rodents is likely to reflect the historical biogeography of the North American desert regions in which they evolved. These rodents are restricted to arid regions, have low vagility, and have a fossil record that documents a history of sufficient antiquity to have evolved along with the developing deserts. Our research has revealed coincident phylogeographic breaks across multiple taxa of the southern warm deserts of North America that coincide (temporally and spatially) with major events in historical geology, and has revealed a number of cryptic forms (e.g., Riddle et al., 2000a, 2000b; Hafner et al., 2008). Of broader import, we have been able to produce hypotheses of regional development of the warm deserts (initially based solely on mammals) that have been tested repeatedly in studies of other taxa of plants and animals (reviewed in Riddle et al., 2000c; Sinclair et al., 2004; Hafner and Riddle, 2005; Lindell et al., 2005; Crews and Hedin, 2006; Devitt, 2006; Riddle and Hafner, 2006a, 2006b; Fehlberg and Ranker, 2007; Wood et al., 2008). In this way, we have compiled a growing list of warm desert taxa that share many of the same historical patterns, leading to an increasingly robust model of the evolution of North American deserts.

One of our early, and now well documented, findings was that the desert biota of the Baja California Peninsula has a unique evolutionary history at least as distinct as that of all other regional deserts (Hafner and Riddle, 1997). This region traditionally has been placed as a subset within the Sonoran Desert, as originally defined by Shreve (1942; Figure 4.1). We proposed recognition of the Peninsular Desert (Hafner and Riddle, 1997) on the same level as the Chihuahuan and Sonoran regional deserts. From a different methodological perspective, Morrone and his colleagues (Arriaga et al., 1997; Morrone, 2001a, 2001b, 2004, 2005) reached the same conclusion: based on phenetic and cladistic studies of a multitude of plant and animal taxa, they recognised the biogeographic province of Baja California as separate from, and at the same level of distinction as, the provinces of Sonora and Altiplano Mexicano, which are equivalent to the Sonoran and Chihuahuan regional deserts.

Our recognition of the Peninsular Desert has not met with wide acceptance, although it is gaining support (e.g., Fehlberg and Ranker, 2007; Leaché et al., 2007). Instead, studies repeatedly figure the Sonoran Desert as including the Baja California Peninsula (e.g., Nason et al., 2002; Hurtado et al., 2004; Sinclair et al., 2004). This lack of general acceptance is not due to a specific debate about the history of the Peninsular Desert; there is a preponderance of support for its evolutionary distinctiveness. Instead, it appears to represent an unwillingness to replace a static, largely physiognomic botanical view of desert regions with one based on shared

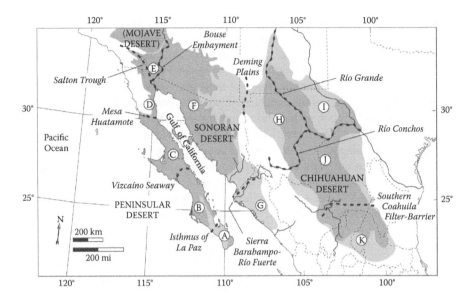

FIGURE 4.1 Warm regional deserts of North America following Shreve (1942, 1951; Sonoran and Chihuahuan; dark shading) and this chapter (Peninsular, Sonoran, and Chihuahuan; light shading). Heavy dashed lines indicate major (putative) vicariant events dividing subregions: A, San Lucan, exclusive of the Sierra Laguna; B, Magdalenan; C, Viscaíno; D, San Felipe; E, Coloradan; F, Sonoran; G, Sinaloan; H, Chihuahuan; I, Trans-Pecos; J, Coahuilan; and K, Zacatecan.

within-region and distinct amongst-region evolutionary history amongst the biotas. The traditional view emphasises phenetic similarity of plant species: in effect, the Peninsular Desert is considered a subset of the Sonoran Desert because of gross similarities in the growth form of the plants. This traditional view effectively ignores a rich and complicated history of diverse geological and climatological impacts on the evolution of the component species of the regional deserts. From an evolutionary perspective, we find this static phenetic view to be of little functional use.

Shreve (1942: p. 215) recognised the dynamic nature of deserts: 'It is not yet possible to state ... to what extent the differences in flora and floristic composition have been influenced by historical factors'. We are now able to describe much of the enormous impact of history that has been hidden beneath the traditional definition of the North American regional deserts. The utility of the 'regional desert' concept would be greatly enhanced by modifying the definition of the regional desert boundaries based on a consideration of evolutionary divergence rather than simple physiognomy, plant community structure, and floristic composition. Phylogeographic analysis of the denizens of these deserts has forced a redrawing of the desert boundaries, provided a greater appreciation of the nature and extent of intermixing amongst the regions, and identified specific filter-barriers within the regional deserts. The vast majority of historical events have been described in an active geological literature, but phylogeographic studies are advancing to the point where we may be able to suggest fruitful areas of new geological research. One splendid example is the hypothesised

Vizcaíno Seaway, a filter-barrier strongly predicted from phylogeographic studies but lacking in detailed geological support.

Herein we review the historical boundaries of the North American warm regional deserts and identify the most informative definition based on taxonomic representation, consideration of major evolutionary relationships, and inclusion of geomorphic and climatic data. We then describe the location, nature, and efficacy of past barriers and current filter-barriers between and within these regional deserts, particularly as they affect the distribution of arid-adapted rodents. Our intent here is to focus on history, considering both the phylogeographic relationships amongst representatives of the different regions and the geographical and ecological barriers and filter-barriers involved in the diversification of species and biotas amongst North American deserts.

4.2 NORTH AMERICAN DESERTS

4.2.1 Warm Regional Deserts

As originally defined by Shreve (1942; Figure 4.1), the North American desert consisted of four subdivisions: the Great Basin, Mojave, Sonoran, and Chihuahuan. Smaller, isolated desert regions were noted in the valleys of Ixmiquilpan, Actópan, Mesquital, and Tehuacán in Hidalgo and Puebla, México. Shreve (1942) made no mention of the San Joaquin Valley desert of California, U.S.A, or the Cuenca Oriental of Puebla and Veracruz, México; two relatively small, isolated, and peripheral desert regions. Axelrod (1983) described the thermal parameters of the western regional deserts, ranging from the cold Great Basin, through the intermediate Mojave, and south to the warm Sonoran Desert. The Mojave Desert is generally considered as a 'warm' desert along with the Sonoran and is included with the Sonoran in the Tropical–Subtropical Desert ecoregion (Bailey, 1997), apart from the Great Basin (Temperate Desert). Elsewhere (Hafner and Riddle, 1997) we advocated separate recognition of a Peninsular regional desert based on the independent evolutionary trajectory of its flora and fauna (vertebrate and invertebrate). Rodents of the Mojave Desert include a mixture of Great Basin, Peninsular, and Sonoran species, as well as several species with a distributional centre in the Mojave (e.g., *Spermophilus mohavensis, Dipodomys panamintinus, D. deserti*). It would be fruitful to evaluate the phylogeographic relationships of the Mojave, Great Basin, and San Joaquin Valley desert species relative to those of the Peninsular and Sonoran regions. Our emphasis here is on the southern warm deserts at approximately the same range of latitudes: Peninsular, Sonoran, and Chihuahuan.

4.2.2 Time Frame and Target Taxa

We initially chose a selected subset of 12 arid-adapted rodent species and species groups (genera *Ammospermophilus, Peromyscus* [subgenus *Haplomylomys*], *Chaetodipus,* and *Dipodomys*) as representative of the warm-desert biota of North America (Riddle et al., 2000c). These rodents are closely tied to arid environments, have low dispersal capability, share the same general distribution in the deserts, and have a fossil history that places them within the developing aridlands of North

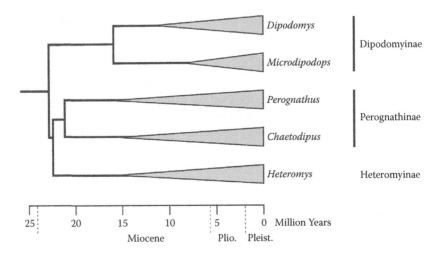

FIGURE 4.2 Evolutionary relationships and timing of species diversification within genera and subfamilies of the rodent family Heteromyidae (after Hafner et al., 2007; *Liomys* is included within the genus *Heteromys*). (After Hafner, J.C., Light, J.A., Hafner, D.J., Hafner, M.S., Reddington, E., Rogers, D.S. and Riddle, B.R. (2007) *Journal of Mammalogy*, 88: 1129–1145.)

America. Thus we hypothesised that the taxa share phylogeographic patterns, particularly in terms of responses to major geological and climatological events. We subsequently expanded the list of taxa to include 25 mammals and (from the literature) 22 birds, six reptiles, one amphibian and one plant species (Riddle and Hafner, 2006a). Additional taxa continue to be added to this growing phylogeographic pattern (e.g., spiders, Crews and Hedin, 2006; rattlesnakes, Castoe et al., 2007; ocotillo and relatives in the family Fouquieriaceae, J. Redfern, pers. comm., 2008). Thirteen of the rodent species are in two subfamilies (Perognathinae and Dipodomyinae) of the family Heteromyidae, which most likely split from its geomyoid sister-taxon, the Geomyidae (pocket gophers), in the Early Oligocene (Hafner et al., 2007; Figure 4.2). Moreover, these two subfamilies have fairly plentiful fossil records that extend back to the Early Miocene. Phylogeographic study of the heteromyids, in particular, thus has the potential to closely correlate major geological events in the evolution of western North America with corresponding lineage divergence and diversification within the family (Alexander and Riddle, 2005; Hafner et al., 2007).

4.2.3 GEOLOGICAL AND PALAEOCLIMATIC EVENTS

The distribution and diversification of the modern regional deserts of North America resulted from a complicated and dynamic geological and climatological history during the Cenozoic (first appreciated and described by Axelrod, 1958, 1983). The origin of these aridlands followed a major period of mountain building during the Early Eocene. This included uplift of the Rocky Mountains and Sierra Madre Oriental, and volcanic origin of the Sierra Madre Occidental (Ortega-Gutiérrez and Guerrero-Garcia, 1982; Swanson and McDowell, 1984; Ferrusquía-Villafranca et al., 2005).

Putative causal events thus far invoked in the subsequent diversification of aridlands taxa range in age from an Early Oligocene palaeoclimatic shift to sea-level changes and range expansion that marked the end of the latest glacial episode (Hafner et al., 2007). The Early Oligocene cooling shift marked an abrupt and extreme shift from mesic to drought-tolerant plant species (Prothero, 1998). A more extreme and rapid cooling shift in the middle Miocene led to the return of permanent ice sheets in the Antarctic, initiated modern ocean circulation patterns, and began the cooling and drying trend that has led to the current glacial episodes (Prothero, 1998; Woodburne, 2004). Tectonic processes then reshaped the western landscape via extensive block-faulting that created the Basin and Range province, uplift of the Cascade and Coast ranges that created additional rain shadows and expelled marine waters from the San Joaquin Valley, resurgent uplift of the Sierra Nevada, intensified plate movement that opened the Gulf of California, and volcanic eruptions along the plate boundary that created the Baja California Peninsula (Norris and Webb, 1976; Gastil et al., 1983; Lonsdale, 1989; Spencer and Normark, 1989; Stock and Hodges, 1989). The Cape Region of the peninsula was originally part of the continental mainland that was transferred to the Pacific plate across the East Pacific Rise, and was later joined to the California mainland as volcanic eruptions and uplift coalesced to form the peninsula. At least a portion of the central peninsula also was transferred across the plate boundary, as evidenced by a terrestrial fossil fauna of Early Eocene age (~53 Ma) from the Tetas de Cabra formation from near Punta Prieta, Baja California Sur on the peninsula (Flynn and Novacek, 1984; Figure 4.3).

4.2.4 NEOGENE GEOMORPHOLOGY AND CLIMATE OF THE BAJA CALIFORNIA PENINSULA

The exact extent, sequence, and relative timing of peninsular formation and of marine inundation at the head of the Gulf of California (Salton Trough and Colorado River) and across the peninsula (Vizcaíno Seaway and Isthmus of La Paz) are the topics of current debate (Figure 4.3; modified from Murphy and Aguirre-León, 2002). As early as 12 Ma (Figure 4.3b), a proto-Gulf formed by a marine basin covered large areas of southeastern California, northwestern Sonoran, and southeastern Arizona, extending north to southern Nevada (Ingle, 1974; Smith, 1991; Gastil et al., 1999; Carreño and Helenes, 2002). It is unclear when the present Gulf opening severed connections between the Sonoran continental mainland and the developing peninsula. Early Eocene fossils of the Punta Prieta vertebrate fauna (Flynn and Novacek, 1984) indicate that the central section of the peninsula was attached to the continental mainland at that time. According to Ledesma-Vázquez (2002), a land connection between the two persisted until the proto-Gulf and current Gulf opening were connected as recently as 3 Ma (shaded areas with question marks in Figures 4.3b–f); whereas Murphy and Aguirre-León (2002) indicate complete separation from the mainland early in the process (Figure 4.3b). Thus the early Blancan Las Tunas local fauna of the Cape Region (Miller, 1980) might represent isolated relicts of continental biota separated by the transfer across plate boundaries 10 Ma or later immigration along a coalesced peninsula via either a Sonoran (mid-peninsular) or Californian (gulf head) continental contact.

The Bouse Formation, which is distributed along the lower Colorado River as far north as southern Nevada, records a brief but deep aquatic inundation variously

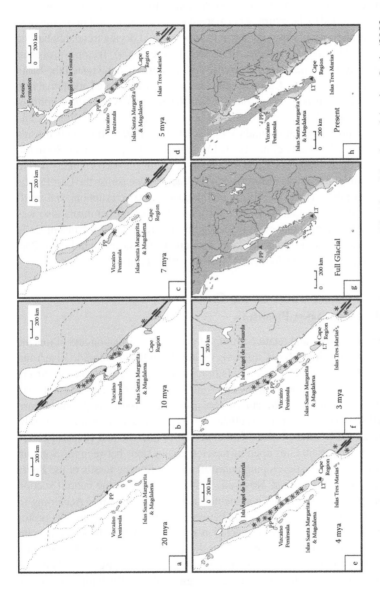

FIGURE 4.3   Summary of the geological evolution of the Baja California Peninsula, modified from Murphy and Aguirre-León (2002; see text). Lighter shading in 4.3g indicates exposed continental shelves (−100 m), darker shading in 3H indicates upland (>100 m). Triangles indicate fossil sites: PP = Early Eocene (~53 Ma) Tetas de Cabra Formation near Punta Prieta, Baja California (Flynn and Novacek, 1984); LT = early Blancan (~4 Ma) Las Tunas local fauna, Baja California Sur. (Modified from Murphy, R.W. and Aguirre-León, G. (2002), in T.J. Case, M.L. Cody and E. Ezcurra (eds.) *A New Island Biogeography of the Sea of Cortéz*. New York: Oxford University Press, pp. 181–220.)

dated as Late Miocene to Late Pliocene in age (Figure 4.3d) and is attributed to either an extension of marine waters from the Gulf or a chain of freshwater lakes (Metzger, 1968; Mattick et al., 1973; Olmsted et al., 1973; Blair, 1978; Eberly and Stanley, 1978; Lucchitta, 1979; Boehm, 1984; Ingle, 1987; Stock and Hodges, 1989; Buising, 1990; Grismer, 1994; Faulds et al., 2001; Spencer and Pearthree, 2001, 2005). Recent studies (summarised by Spencer and Pearthree, 2005: p. 2) indicate that the Bouse Formation probably resulted from 'a series of lakes that filled with river water and spilled over, eventually linking the river with the Gulf of California'. It appears that a through-flowing Colorado River developed by 5.3 Ma, simultaneously cutting the Grand Canyon and filling the mouth of the Salton Trough with sediment. By 4 Ma, this sediment had isolated the Salton Trough from the Gulf.

There is as yet no geological evidence for a Pliocene-age seaway connecting the Gulf with an inundated Los Angeles Basin (Figure 4.3d), as suggested by Murphy (1983), Boehm (1984, 1987), and Murphy and Aguirre-León (2002). Although marine inundation of the Salton Trough and Los Angeles Basin would probably have resulted in effective isolation of many peninsular lineages from their North American mainland counterparts (Grismer, 1994), an incomplete filter-barrier would not fully explain disjunction in both mesic-adapted and xeric-adapted lineages (Murphy and Aguirre-León, 2002). The nature and timing of perhaps multiple inundations along both the Salton Trough and Colorado River is complicated by subsidence (via broad regional downwarping), uplift of the Colorado Plateau, and shifts along the southernmost section of the tectonically active San Andreas Fault and a complex pattern of active ridge segments and transform faults that continues beneath the Gulf (Buising, 1990; Petersen and Wesnousky, 1994; Spencer and Pearthree, 2005).

Geological and palaeontological evidence support a Late Miocene or Early Pliocene submergence of the central peninsula, in the vicinity of the Vizcaíno Desert, associated with the opening of the initial proto-Gulf (Figure 4.3c,d; Helenes and Carreño, 1999; Carreño and Helenes, 2002; Ledesma-Vázquez, 2002). Indeed, fossil marine molluscs document the submergence during this time of huge areas of the peninsula, including Pacific waters extending to San Ignacio on the eastern edge of the Vizcaíno (Smith, 1984, 1991) and marine sedimentation beginning around 7 Ma on the Gulf coast near Santa Rosalía (Ortlieb, 1991; Holt et al., 2000). The extensive molecular sequence divergence observed amongst a variety of terrestrial and marine lineages in the area support such an early date for a Vizcaíno Seaway (Murphy and Aguirre-León, 2002; e.g., reptiles, Rodríguez-Robles and De Jesús-Escobar, 2000; Lindell et al., 2005; spiders, Crews and Hedin, 2006; near-shore reef fishes, Riginos, 2005).

At the same time, other lineages showing congruent genealogical breaks in or around the mid-peninsular region are characterised by more subtle levels of sequence divergence (Riddle et al., 2000c; Riddle and Hafner, 2006a), arguing perhaps for a more recent vicariant event (such as the initial 1-Ma date for a mid-peninsular seaway suggested by Upton and Murphy, 1997) and inviting Grismer (2002) to speculate that these genetic divergences instead may have resulted from abrupt ecological and climatic changes in the central peninsula. Although a 'vicariant event affecting all species similarly' remains the most parsimonious explanation for the large number of co-distributed lineages (nearly 20 terrestrial and 10 marine) in the mid-peninsular region (Lindell et al., 2006: p. 1329), could multiple, more recent transpeninsular seaways have existed in the Vizcaíno region (Figures 4.3C–F), as suggested by Lindell et al. (2005) and Crews and Hedin (2006)? The Vizcaíno Desert of the central Baja California Peninsula is a flat expanse between the twin lagunas of Ojo Liebre (Scammons) and San Ignacio (Figure 4.4). The low-lying region connecting the two lagunas is dotted with a series of dry lakes (Camacho, Huaca, Grande, and Palvadanes) and reaches a maximum elevation of only 30 m. Limestone exposed along the peninsular highway may have been deposited in shallow warm seas during the last 1 Ma (Minch and Leslie, 1991), but they are virtually fossil-free, lacking even foraminiferan (J. Minch, pers. comm., 1999). The desert is bordered on the west by ancient volcanic uplands of the Vizcaíno Peninsula and on the east by the Palaeogene-age volcanic spine of the peninsula. This spine is interrupted between the Sierra San Francisco (the southern end of the northern Peninsular Ranges basement) and the northern extension of the Sierra de la Giganta (the southern half of the

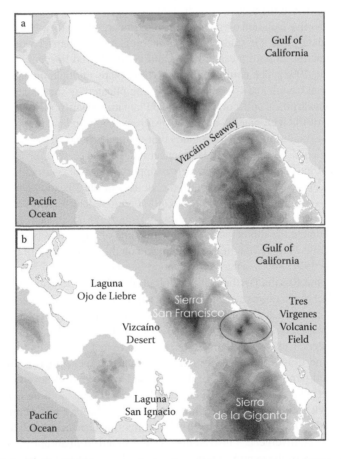

FIGURE 4.4   (a) Possible latest transpeninsular seaway across the central Baja California Peninsula, prior to the eruption of the Tres Virgenes lava field and associated uplift of the Vizcaíno Desert ~1 Ma. (b) Current topography of the Vizcaíno region of the central Baja California Peninsula. Contour interval = 100 m.

Peninsular Ranges; Figure 4.4) by a narrow canyon that follows transform faults (and along which the peninsular highway passes). The gap between the two Palaeogene-age ranges was plugged by three Quaternary volcanic eruptions: the La Reforma caldera on the Gulf coast (1.2 Ma: Schmidt, 1975; Sawlan, 1986); the El Aguajito caldera (0.8 Ma; Garduño-Monroy et al., 1993); and the westernmost Tres Virgenes volcanic complex, beginning with El Viejo, followed by El Azufre, and culminating with Volcán La Virgen, which began erupting 26 Ka and now reaches 1920 m.

Although the last eruptive phase of La Virgen is dated at 6515 yBP (Capra et al., 1998), a Jesuit missionary, Ferdinando Consag, reported an eruption in this area in 1746 (Ives, 1962), steaming was witnessed in 1857 (Dickey, 1983), and there appears to be strong potential for future eruptions (Schmitt et al., 2006; it is the current site of a geothermal energy development). The extent of uplift in the Vizcaíno region associated with the Quaternary volcanics, particularly the Tres Virgenes complex, is unknown, although extensive uplift of marine sediments has been documented along the eastern edge of the volcanic field north of Santa Rosalía (Anderson, 1950; Durham and Allison, 1960). Moreover, marine coquinas have been found beneath the oldest La Reforma flow deposits, indicating that marine waters of the Gulf occupied this region before eruption of La Reforma (Hafner and Riddle, 2005; B. Hausback, pers. comm., 2000). However, older (several million years old) volcanic deposits surrounding San Ignacio are subaerial and do not appear to have been submerged for any length of time (i.e., no overlying marine deposits; B. Hausback, pers. comm., 2000). It is possible that a seaway, or periodic seaways, existed at the latitude of San Ignacio (27°20′N) up until about 1 Ma, or that a very narrow landbridge existed between a flooded Vizcaíno Desert and the Gulf, with extensive volcanic eruptions occurring on the narrow landbridge (Hafner and Riddle, 2005).

The twin lagunas of Ojo Liebre and San Ignacio might have been joined as recently as 6000 yBP (Xerithermal) by a modest 5–10 m rise in sea level if the current elevation of the Vizcaíno floor was achieved by subsequent uplift related to eruption of Volcán La Virgen. This would have isolated the Vizcaíno Peninsula as an island, isolation that may be recorded in molecular divergence amongst lineages of the western and eastern margins of the Vizcaíno Desert. Leaché et al. (2007) tested for simultaneous vicariance of co-distributed taxa of mammals and reptiles across the Vizcaíno phylogeography break, employing a Bayesian analysis of an hierarchical model, and instead found support for two events. The range of sequence divergences (from extensive to subtle) between sets of co-distributed taxa at or near the Vizcaíno Desert, coupled with the abrupt ecological and climatic shifts in the vicinity, indicates a far greater complexity of historical explanation than has thus far been offered and highlights the need for additional and more detailed geological and palaeontological study.

The Cape Region of the Baja California Peninsula and its relict mainland flora and fauna may have experienced repeated joining with and disjunction from the peninsula due to marine flooding of the San José trough (between the Sierra Santa Clara and Sierra Laguna) and across the narrow, low-elevation Isthmus of La Paz immediately west of La Paz (Figure 4.3F; McCloy, 1984). The as-yet unknown affinities of elements of the early Blancan Las Tunas local fauna (Miller, 1980) may lie with the Trans-Mexican Volcanic Belt along Jalisco, the source of the initial transfer of the

Cape Region across the plate boundary 10 Ma, or more recent (4 Ma) contact with either mainland Sonora (to the east) or California (north). A broader reconnaissance of possible Pliocene and Pleistocene terrestrial and marine deposits coupled with a more exact understanding of the history of uplift, subsidence, and fault dynamics in this region will be required to better evaluate patterns of genetic divergence observed in Cape Region lineages.

Repeated climatic cycles of the late Pleistocene, particularly of the last 700 Ky (Webb and Bartlein, 1992), caused sea-level fluctuations and shifting of ecological zones (in latitude and elevation; summarised in Hafner and Riddle, 2005). Lowered sea levels during pluvial maxima (66–122 m; Durham and Allison, 1960; Rosen, 1978) exposed vast areas of continental shelves (particularly along the southwestern edge of the Baja California Peninsula), most of the shallow gulf north of San Felipe (~31°N), and landbridges to many of the near-shore islands surrounding the peninsula (Figure 4.3G). Exposed shelves and landbridges expanded habitat and provided dispersal corridors, particularly for sand-dwelling taxa. Increased moisture and lowered temperatures allowed woodland to extend lower in elevation and latitude: the presence of disjunct mesic-adapted mammals in the Cape Region (*Peromyscus truei* and *Sorex ornatus*) and fossil evidence (Miller, 1977; Rhode, 2002) indicate more continuous and denser grassland and woodland cover along the peninsula during pluvial maxima. During the Pleistocene, the Salton Trough alternated between dry playa and dunes 84 m below sea level, freshwater lakes (the largest, Lake Cahuilla [= Leconte], some six times the size of the present Salton Sea; Figure 4.3G), and highly saline remnants of evaporating lakes. Filling of the Salton Trough to form Lake Cahuilla (Thompson, 1968; Van de Kamp, 1973) severely restricted biotic communication between the southern-depressed arid-adapted species of the peninsula and neighbouring relatives in the Mojave and Sonoran regions.

4.2.5   NEOGENE GEOMORPHOLOGY AND CLIMATE OF THE SONORAN AND CHIHUAHUAN REGIONAL DESERTS

Pliocene and Pleistocene geomorphological processes associated with filter-barriers elsewhere in the North American regional deserts appear to be less complicated than those of the Baja California Peninsula. The major landforms of the region were in place: the Sierra Madre Oriental had formed in the early Palaeogene (along with the Rocky Mountains) from folding and thrusting of Cretaceous and Jurassic limestone (Ortega-Gutiérrez and Guerrero-Garcia, 1982; Ferrusquía-Villafranca et al., 2005), and intensified volcanic activity associated with increased lithospheric subduction between 34 and 23 Ma produced the gigantic volcanic field of the Sierra Madre Occidental and smaller associated fields in southwestern New Mexico (Mogollon–Datil plateau, Black Range and Peloncillo Mountains). Subsequent block faulting and erosion broke and wore away the eastern and western fringes of the Sierra Madre Occidental, developing the series of major rivers that drain the western watersheds and forming extensive basins in the elevated Mexican Plateau. Block faulting (continuing today) in southwestern New Mexico reshaped the area in the vicinity of today's Deming Plains and rift-related block faulting, headward erosion, and stream capture redirected the flow of the Gila River, Río Grande, and Pecos River. There were three major filter-barriers in the combined Sonoran–Chihuahuan region: (1) the Cochise filter-barrier (Morafka, 1977a, 1977b) in the vicinity of the Deming Plains, traditionally considered as the point of contact between the two regional deserts; (2) the Southern Coahuila filter-barrier (Baker, 1956; Baker and Greer, 1962; Peterson, 1976; Schmidly, 1977); and (3) the Sierra Barabampo–Río Fuerte filter-barrier (Hafner and Riddle, 2005).

The Cochise filter-barrier is located in the Mexican Highland section of the Basin and Range province of southeastern Arizona and southwestern New Mexico. This region extends from the Baboquivari Mountains of south-central Arizona east across the Peloncillo Mountains and onto the elevated Deming Plains, east to the Río Grande rift and the Río Grande, and south to the Río Conchos. The vicinity of the Deming Plains marks a relatively low-elevation gap in the Continental Divide along the Sierra Madre Occidental that continues north into the mountains of western New Mexico. In fact, it is a large, 200-km wide closed basin (Southwest Divide basin; Hunt, 1983) that stretches from the Arizona–New Mexico border to the western escarpment of the Río Grande and was the former sump of the Gila River, Mimbres River, and Río Grande. During the Late Miocene and Early Pliocene, flow from the Gila and Mimbres rivers maintained extensive lakes in the basin; Hawley et al. (2000) argue for this older age for the lakes, contra the middle Pleistocene age model advocated by Morrison (1965, 1985, 1991). By the Pliocene, the Río Grande had managed to develop a through-flowing system via closed-basin spillover along the basins of the Río Grande rift (Mack et al., 1998; Pazzaglia and Hawley, 2004; Connell et al., 2005) and added its flow to the closed Southwest Divide basin lake system (Figure 4.3F).

About 2 Ma the rift basins of southern New Mexico filled to form Lake Cabeza de Vaca (Strain, 1966, 1971; Campos-Enriquez et al., 1999). Subsequently, drainage shifted toward the south, forming Laguna Palomas in northern Chihuahua (Campos-Enriquez et al., 1999). Near the Plio-Pleistocene border, headward erosion eventually captured the ancestral headwaters of the Gila River, which was integrated downstream and redirected to westward flow into Arizona and eventually into the Colorado River. At approximately the same time (~2.3 Ma; Connell et al., 2005), the Río Grande was diverted southward through El Paso, Texas, 'either by capture by a tributary of the Río Conchos or by overflow at a meander located against the El Paso Gap' (Hunt, 1983: p. 42), joined with the Río Conchos and emptied into the Gulf of Mexico. The combined Río Grande–Río Conchos flow was increased in the early Pleistocene when a tributary of the ancestral Río Grande cut across headwaters of east-flowing rivers draining the Llano Estacado of eastern New Mexico (including the ancestral Pecos River), forming the current Pecos River drainage. Climatic oscillations of the Pleistocene alternated in opening (during interpluvials) and closing (during pluvials) the Deming Plains dispersal corridor for arid-adapted species. During pluvial periods (Figure 4.3G), the closed basins in southern New Mexico, northern Chihuahua, and Trans-Pecos Texas filled with lakes (Smith and Street-Perrott, 1983; Wilkins and Currey, 1997; Campos-Enriquez et al., 1999), deserts were supplanted by more mesic-adapted vegetation, and riparian vegetation corridors broadened and integrated. During interpluvials, lakes dried to playas and desert species reinvaded and swept across the gap in the Continental Divide.

The Southern Coahuila filter-barrier is composed of three components, from west to east: (1) the Río Nazas (Baker and Greer, 1962) and Río Aguanaval (Hafner et al., 2008), which drain eastern watersheds of the Sierra Madre Oriental and empty into the closed terminal bolson of the Laguna Mayrán; (2) the Laguna (Desierto) Mayrán itself, which has been a playa throughout the Holocene (K. Butzer, pers. comm., 2005); and (3) a western extension of the Sierra Madre Oriental, including the Sierra de Parras and Sierra de Guadalupe, intermittently broken by lower eleva- tion gaps. These western rivers are both subject to catastrophic floods induced by hurricane-related precipitation events that inundate much of the Río Nazas and Río Aguanaval alluvial plains. Following such events, open water may persist in regions of the Mayrán Basin for several years, and it is likely that higher and more constant water levels existed during pluvial periods of the Pleistocene. Similarly, pluvial cli- mate would have lowered montane vegetative zones and expanded riparian corri- dors, enhancing the severity of the filter-barrier to arid-adapted biota.

The Sierra Barabampo and Río Fuerte run parallel along the Sonora–Sinaloa bor- der from the Sierra Madre Occidental to the Gulf of California, and together form a filter-barrier to flora and fauna of the coastal plains. It is the dominant feature of these monotonous plains from the Sierra Tinaja del Carmen (28°N) to Cerro La Punta in southern Sinaloa (22°30′N), and has probably been the dominant filter- barrier for those taxa that were able to cross the other major rivers that cut across the coastal plains. The Sierra Barabampo, a perpendicular extension of the Sierra Madre Occidental, is composed of volcanic extrusives of the Paleocene to Oligocene age. It rises abruptly above the coastal plains and plunges into the gulf, forming the bulging cape above Los Mochis and Topolobampo. Today the major north–south highway climbs through a narrow, steep-walled canyon that pierces the sierra. The wide chan- nel of the Río Fuerte attests to its prodigious flow prior to diversion for agriculture and construction of hydroelectric dams. It is fed by rivers of the six major canyons of the 6.5 Mha Barrancas del Cobre drainage, four canyons of which are deeper than the Grand Canyon of the Colorado River. During pluvial maxima, flow of the Río Fuerte undoubtedly increased, and montane vegetation crept down the slopes of the Sierra Madre Occidental, increasing the severity of the filter-barrier.

4.2.6 TEMPORALLY NESTED EVENTS

It is important to note that, for virtually every filter-barrier described herein, there exist multiple, temporally nested events that would have resulted in similar vicari- ance or dispersal events across diverse temporal scales (e.g., Miocene to last glacial maximum). Moreover, the more detailed the knowledge gained about the history of each region, the more temporally complicated that history has appeared. Multiple mid-peninsular seaways across the Baja California Peninsula in the vicinity of the Vizcaíno Desert may have existed between 7 Ma and 1 Ma, and partial flooding of the low-lying desert during interpluvials (particularly the Xerithermal, 6 Ka) and present-day ecological shifts in the vicinity could echo the effects of more ancient vicariant events. At the head of the Gulf of California, the effects of pluvial Lake Cahuilla and the Colorado River could have effected similar but less trenchant patterns of genetic divergence as Pliocene and Miocene freshwater lakes (Bouse

Formation) and marine transgressions along the Salton Trough and along the current bed of the Colorado River. The Cochise filter-barrier had its birth in the initial Oligocene–Miocene eruption of the Sierra Madre Occidental volcanic field, experienced landscape transformation and shifts in river flow during Pliocene and Pleistocene block-faulting, and was repeatedly opened and shut by vegetative shifts during Pleistocene climatic oscillations. The Sierra Barabampo–Río Fuerte and Southern Coahuila filter-barriers were probably more stable through time, yet were subjected to the same repeated climatic shifts of the Pleistocene.

4.3   DISTRIBUTION OF ARID-ADAPTED RODENTS

We have mapped the combined distribution of our selected subset of 31 evolutionarily significant units (ESUs, *sensu* Moritz et al., 1995) of arid-adapted rodents of the southern, warm regional deserts (Figure 4.5; distributions generally from Hall, 1981, but modified by subsequent records). We have magnified the geographic separation of allopatric species or ESUs in order to illustrate filter-barriers at this geographic scale, as opposed to undifferentiated species of continuous distribution. In this depiction, there are two quite distinct sources of high species density: core areas of endemism and areas of sympatry between neighbouring areas of endemism. Discrimination between the two sources of species density is provided by graphical analyses of the exact nature of species turnover at filter-barriers (see below). However, this depiction illustrates the blurry nature of boundaries of the regional deserts, depending on the idiosyncratic ecological limitations and dispersal capabilities of each species, while indicating three warm regional deserts: Peninsular, Sonoran, and Chihuahuan.

4.4   REGIONAL DESERT BOUNDARIES

Shreve (1942) defined the North American regional deserts based wholly on characteristics of their plants: physiognomy, community structure, and floristic composition. Shreve (1942: p. 211) specifically stated that these deserts are "… like any other great natural region in being without sharp boundaries …" but emphasised the relative uniformity within each region and considered only community succession as a historical process. A variety of authors has suggested alternative boundaries based usually on climate or flora. For example, Schmidt (1989) reviewed 17 different definitions of the Sonoran Desert alone, with a southern boundary on the Sonoran–Sinaloan mainland variously placed between 27°30′N and 22°N, a distance of >750 km.

Arriaga et al. (1997) defined the boundaries of the warm deserts in Mexico based on a combination of four different biogeographic systems: (1) morphotectonic features (Ferrusquía-Villafranca, 1990); (2) flora (Rzedowski and Reyna-Trujillo, 1990); (3) herpetofauna (Casas-Andreu and Reyna-Trujillo, 1990); and (4) mammalian fauna (Ramírez-Pulido and Castro-Campillo, 1990). This arrangement, subsequently slightly modified by Morrone (2001b, 2004, 2005) recognised three arid 'provinces': Baja California (corresponding to our Peninsular regional desert), Sonoran, and Altiplano Mexicano (= Chihuahuan regional desert). Morrone (2005) added an important historical element by identifying panbiogeographic generalised tracks amongst these regions, augmenting the already extensive database with

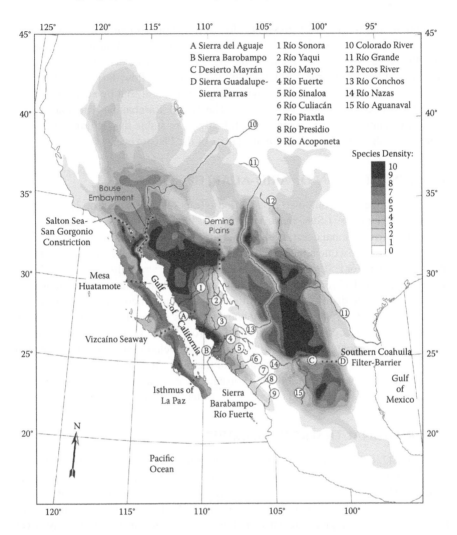

FIGURE 4.5 Contour map of species density based on the combined distribution of 31 Evolutionarily Significant Units (ESUs) of arid-adapted rodents. Separation of ESUs at points of allopatric contact is exaggerated to be visible at this geographic scale.

studies of insects, fish, lizards, and plants (Marshall and Liebherr, 2000), birds (Zink et al., 2000), and beetles (Morrone and Márquez, 2001). Riddle and Hafner (2006b) conducted an independent cladistic analysis of 22 clades of arid-adapted animals and plants using an entirely different methodological approach (Brooks Parsimony Analysis, BPA; Brooks et al., 2001). Generalised tracks (Morrone, 2005) and BPA (Riddle and Hafner, 2006a) strongly indicate that the Baja California Peninsular region is either the outgroup to, or at least equivalent in divergence to, the Sonoran and Chihuahuan regions. Moreover, the boundaries shown for the three regions by Arriaga et al. (1997) and Morrone (2005) are remarkably similar to those indicated by our selected subset of arid-adapted rodents.

4.4.1 EMBEDDED SUBREGIONS

We have adopted an explicitly historical approach to examining possible subregions embedded within the three southern, warm deserts. We first identified putative vicariant events and filter-barriers (past or present) that might have effected differentiation within and amongst regions (Figure 4.1), and then examined the phylogeographic structure of our selected subset of desert rodents. We reasoned that past vicariant events might have resulted in speciation of some lineages and not others, such that species-level analysis would be inadequate to identify common phylogeographic patterns. In contrast, some record of past vicariance (if it existed) should have been retained in the mitochondrial DNA (mtDNA) of the affected taxon. As expected, we found uniform distribution of mtDNA haplotypes within areas defined by putative filter-barriers. We also discovered marked differentiation of mtDNA haplotypes coincident with the location of past vicariant events and putative filter-barriers. In several cases, this has revealed cryptic species that had been hidden by shared morphological appearance (Riddle et al., 2000a, 2000b).

For purposes of reconstructing the history of biotic diversification in separate areas that also exhibit an history of admixture following isolation and divergence, it is important to distinguish between core areas of endemism, peripheral depauperate areas and areas of faunal intermixing. One objective method that we have used is parsimony analysis of endemicity (PAE; Rosen, 1988). PAE was originally intended to infer historical area relationships, but has not been considered an historical method because it fails to employ taxon phylogeny information (Humphries and Parenti, 1999). Morrone (1994) and Morrone and Crisci (1995) have demonstrated that PAE is useful in identifying provisional areas of endemism. In our previous analysis (Riddle and Hafner, 2006a), we began with eleven candidate areas of endemism within the regional deserts (and five peripheral areas) based on the distribution of 22 clades and 10 putative vicariant events or filter-barriers. PAE identified an eastern region composed of three areas (Continental East) that corresponds to the Chihuahuan Desert, a centrally-located Continental West region (two areas) corresponding to the Sonoran Desert, and an outgroup consisting of two regions (Peninsular North and South) that together correspond to the Peninsular Desert (Figure 4.6). Three areas between these major units could not be assigned to one or another regional desert: the San Felipe and Coloradan areas at the head of the Gulf of California, and the Chihuahuan area (southern New Mexico and Chihuahua north of the Río Conchos). We consider in turn (below) the extent of the North American deserts; the nature and location of the boundaries between the Peninsular, Sonoran, and Chihuahuan deserts; and subdivisions within each of the regional deserts. These borders and barriers are based on phylogeographic analysis of our selected subset of arid-adapted rodents and general patterns observed by Arriaga et al. (1997) and Morrone (2005), as well as consideration of putative causal geological and climatological events.

4.4.2 EXTENT OF NORTH AMERICAN DESERTS

A review of maps depicting the North American deserts (e.g., Shreve, 1942; Wauer and Riskind, 1977) reveals general agreement regarding the full extent of the major

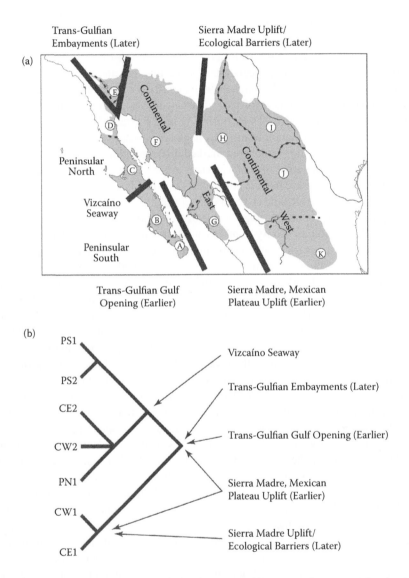

FIGURE 4.6   Depiction of postulated vicariant events (a) amongst four areas of endemism, and (b) a model of historical vicariance as the events are associated with a secondary BPA tree. ([b] After Riddle, B.R. and Hafner, D.J. (2006a). *Journal of Arid Environments*, 66: 435–461.)

arid regions, with the notable exception of two regions: the transition between the Sonoran and Chihuahuan deserts (junction of states of Arizona, New Mexico, Chihuahua, and Sonora) and the southern edge of the Sonoran Desert. Smaller peripheral deserts that are often omitted from these maps include the San Joaquin Valley of California, and the southern peripheral deserts of Mexico, sometimes referred to collectively as the Hidalgan Desert. This latter region extends from southern San Luis Potosí to northern Oaxaca, and includes the Valle del Mezquital, Valle de Ixmiquilpan, Valle de Actopan (Hidalgo), Cuenca Oriental (Veracruz and Puebla), and Valle de Tehuacán (Puebla and Oaxaca). We are concerned here with the boundaries and subunits of the major, southern warm deserts, which we consider to be the Peninsula, Sonoran, and Chihuahuan regional deserts.

4.4.3 PENINSULAR–SONORAN BOUNDARY

Numerous lines of evidence now support the evolutionary distinction of the Peninsular from the Sonoran Desert and further indicate that the Peninsular Desert established an independent evolutionary trajectory before the separation of the Sonoran and Chihuahuan regional deserts. Phylogeographic analysis of aridlands rodents reveals a wide transitional area (Figure 4.7) between the two regions that stretches around the Gulf of California from the Colorado River to the Mesa Huatamote (30°N; Figure 4.1). This transitional area, commonly referred to as the Colorado or San Felipe Desert, is complicated by the overlap of species from a third regional desert, the Mojave. As such, this most arid part of the North American deserts, which

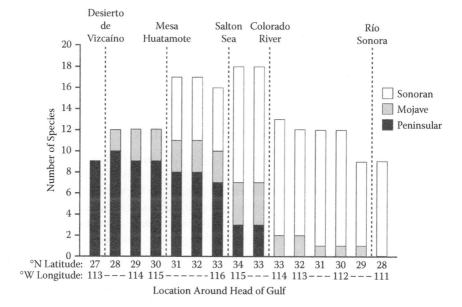

FIGURE 4.7 Shift in proportion of 10 Peninsular, 4 Mojave, and 11 Sonoran species of arid-adapted rodents moving from northern Baja California around the head of the Gulf of California and south into Sonora.

is peripheral to all three regional deserts, currently possesses the highest species diversity of aridlands rodents. For example, five distinct species of pocket mice (*Chaetodipus*) occur in the San Felipe Desert and two other species occur in the nearby foothills or peninsular mountains. Most North American desert regions are characterised by two or occasionally three species of these pocket mice.

It is likely that a variety of extant species in the southern Baja California Peninsula are relicts of the initial separation of the Cape Region from the Mexican mainland 7–10 Ma. These include plants (Roberts, 1989), herptiles (Murphy, 1983; Grismer, 2002; Murphy and Aguirre-León, 2002), birds (Cody, 1983; Cody and Velarde, 2002), insects (Truxal, 1960), spiders (Chamberlin, 1924), and scorpions (Williams, 1980; Gantenbein et al., 2001; Sissom and Hendrixson, 2005). In contrast, there is no direct (e.g., fossil) evidence that any extant mammals of the Baja California Peninsula are relicts of this initial separation (Hafner and Riddle, 2005), even if early Blancan fossils in the Las Tunas local fauna (Miller, 1980) were present at the initial transfer across the plate boundary (see above discussion). Arid-adapted rodents could have gained access to the peninsula soon after connection of the Cape Region with the California mainland by intervening volcanic eruptions and uplift along the fault boundary, or directly from the Sonoran mainland prior to the formation of the modern Gulf opening (via landbridges indicated by Ledesma-Vázquez, 2002). The connection between the peninsula and the mainland may have been severed repeatedly by northern marine inundations up the Salton Trough and up the course of the Colorado River, a system of closed-basin spillover freshwater lakes (Spencer and Pearthree, 2001), and Pleistocene pluvial Lake Cahuilla. These freshwater, marine, or estuarine barriers effectively divided the desert rodent fauna into Peninsular, Mojave, and Sonoran clades. The inhospitable playa and dunes of the Salton Trough and the through-flowing Colorado River have persisted as filter-barriers. The Mesa Huatamote, a southeastern extension of the Sierra San Pedro Mártir, limits dispersal between the San Felipe transitional region and the Peninsular Desert. Steep rocky cliffs mark the edge of the mesa at the Gulf of California overlooking Isla Miramar (= Isla El Muerto; 30°N).

4.4.4 Extent and Subdivisions of the Peninsular Desert

Shreve (1942) excluded from his Sonoran Desert (in which he included the Baja California Peninsula) the thorn forests and arid tropical forests of the Cape Region (south of 24°N on the peninsula). The Sierra Laguna of the Cape Region supports an endemic insular flora and fauna that reflect its long history as an oceanic island ripped from the continental mainland (see above). Accordingly, many floral and faunal elements have their closest relatives on the subtropical mainland and certainly the higher reaches and foothills of the Sierra Laguna do not support desert flora and fauna. Similar 'sky islands' of montane environment are scattered throughout the North American deserts and, like these others, the Sierra Laguna is surrounded by warm desert scrub vegetation (in this case, the Peninsular Desert) at lower elevations.

The Peninsular Desert biota is strongly divided across multiple plant and animal taxa into northern and southern components at approximately the Vizcaíno Desert. A smaller range of taxa, particularly herptiles (Grismer, 1994) and plants (Roberts,

1989), demonstrate extensive endemism in the Cape Region biota, and classical botanical studies (e.g., Shreve and Wiggins, 1964) wholly excluded the Cape Region from regional deserts. There are clearly two components of the Cape Region biota: ancient relicts of the initial transfer of a mainland landmass to the Pacific Plate and subsequent invaders. It remains to be seen if there is a third component: more recent endemic species that invaded the Cape Region following its junction with the new Peninsular mainland, were subsequently isolated by flooding of the San José Trough or the Isthmus of La Paz, and diverged in isolation.

4.4.5 SONORAN–CHIHUAHUAN BOUNDARY

Shreve (1942: p. 213) considered the Sonoran and Chihuahuan deserts to be divided by a gap of 'Desert-Grassland Transition' in the highlands that include the Continental Divide, from the Baboquivari Mountains of Arizona (112°W) to the eastern edge of the Deming Plains of New Mexico (108°W). Brown and Lomolino (1998) narrowed the gap (110–112°W) between the regional deserts, extending the Chihuahuan Desert to the west. Others eliminated the gap entirely, but the line of transition has bounced from 110°W (Brown and Kurzius, 1987) to 112°W (Morrone, 2001a, 2004, 2005) to 109°W (Hafner and Riddle, 2005). Examination of the phylogeographic structure of our selected subset of arid-adapted rodents indicates that these various borders are within a broad zone of transition and overlap between species of the two regions. Moreover, the eastern edge of this transitional area extends far beyond the Deming Plains to the Río Grande and south to the Río Conchos. Dominance of Sonoran rodents shifts to dominance of Chihuahuan rodents across the Deming Plains (108–109°W; Figure 4.8). Although the initial divergence of perhaps many Sonoran and Chihuahuan lineages dates to the initial formation of the Sierra Madre Occidental (location of the Cochise filter-barrier, along the Arizona–New Mexico border), it is likely that the region from the Baboquivari Mountains of Arizona to Trans-Pecos Texas and south to the Río Conchos of central Chihuahua has experienced repeated dispersal east of Sonoran elements and north and west from Chihuahuan elements, and should be considered as transitional in character. Thus the heart of the Sonoran Desert ends in south-central Arizona, and that of the Chihuahuan Desert begins south of the Río Conchos, actually excluding most of the regional desert of Chihuahua (and what we have previously called the Chihuahuan subregion of the Chihuahuan Desert; Figure 4.1; Hafner and Riddle, 2005).

At the Cochise filter-barrier, a combination of macrofossils from packrat (*Neotoma*) middens and bioclimatic envelopes were used by Holmgren et al. (2007) to document shifting distributions of C4 grasslands and C3 shrublands across the zone during the last glacial–interglacial cycle. Ongoing invasion of desert shrubs into former grassland in this area in response to recent climatic change (Brown et al., 1997) is supported by the recent eastward expansion of a desert shrub, jojoba (*Simmondsia chinensis*; Holmgren et al., 2007), and replacement of a grassland-adapted pocket mouse (*Chaetodipus hispidus*) by *C. baileyi* at the extreme eastern periphery of the latter species' distribution (D. Hafner, pers. obs., 2003). The distributions of jojoba and *C. baileyi* have been either causally (Sherbrooke 1976) or spuriously linked (M'Closkey 1983).

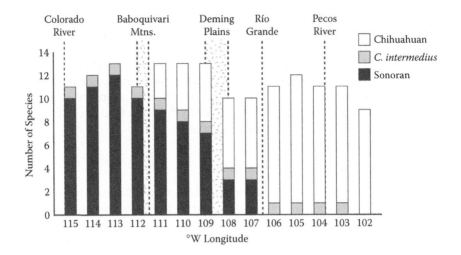

FIGURE 4.8   Shift in proportion of 12 Sonoran and 10 Chihuahuan species of arid-adapted rodents across the southwestern United States between latitudes 31°N and 32°N. The pocket mouse *Chaetodipus intermedius* occurs in the Sonoran subregion of the Sonoran Desert and the Chihuahuan and Trans-Pecos subregions of the Chihuahuan Desert (see Figure 4.1).

4.4.6   SOUTHERN EXTENT AND SUBDIVISIONS OF THE SONORAN DESERT

Shreve's (1942) depiction of the southern extent of the Sonoran Desert (in which he included the Sonoran mainland and the Baja California Peninsula) was conservative. He excluded from desert regions the thorn forests and arid tropical forests of the Cape Region of the Baja California Peninsula (24°N) and of the Sonoran–Sinaloan coastal plains south of the Río Yaqui (27°N; Figure 4.1). As noted earlier, the Sierra Laguna of the Cape Region supports an endemic insular flora and fauna that reflect its long history as an oceanic island ripped from the continental mainland. Accordingly, many floral and faunal elements have their closest relatives on the subtropical mainland.

The postulated southern margin of the continental portion of the Sonoran Desert (to which we restrict the Sonoran regional desert) has shifted back and forth by at least 750 km (Thornwaite, 1931; Smith, 1941; Shreve, 1942, 1951; Dunbier, 1968; Instituto de Geografía, 1970; Rzedowski, 1973; MacMahon, 1979; Brown, 1982; Garcia and de Byres, 1984; Schmidt, 1989; Arriaga et al., 1997; Brown and Lomolino, 1998; Hafner and Riddle, 2005; Morrone, 2005). Major rivers that drain the Sierra Madre Occidental cross the coastal plains of Sonora and Sinaloa at regular intervals (Figure 4.5), and these rivers act as a series of hurdles to species of the coastal plains. The southern margin for the Sonoran Desert has been variously set as far north as the Río Mayo in Sonora (27°30′N; Shreve, 1942, 1951; Rzedowski, 1973; Brown, 1982) and as far south as the Río Acoponeta on the Sinaloa-Nayarit border (22°N; Rzedowski, 1978; Hafner and Riddle, 2005). The most comprehensive of these studies (Arriaga et al., 1997; Morrone, 2001b, 2004, 2005) depict the southern margin at the Río Piaxtla, just north of the Tropic of Cancer (23°27′N). Most of these studies either fixed the southern margin or indicated a subregional shift at or near the Sierra Barabampo and

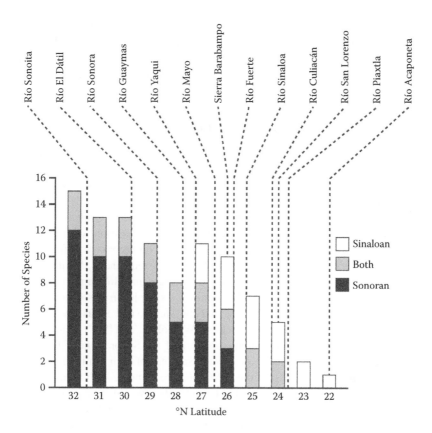

FIGURE 4.9 Shift in proportion of arid-adapted rodents of the Sonoran (12 spp.) and Sinaloan (4 spp.) subregions from northern Sonora through Sinaloa to northern Nayarit, México. Three species occur in both subregions.

Río Fuerte. Our analysis of rodent phylogeography concurs with Arriaga et al. (1997) and Morrone (2001b, 2004, 2005) in including the Sinaloan thornscrub within the Sonoran Desert, and further recognises a shift from the Sonoran subregion to the Sinaloan subregion at the Sierra Barabampo–Río Fuerte (Figure 4.9).

The role of the multiple rivers as a series of filter-barriers along the Sonoran–Sinaloan coastal plains and the dominant role of the Sierra Barabampo–Río Fuerte filter-barrier are perhaps best demonstrated in one species of pocket mouse, *Chaetodipus goldmani*. This species occupies thornscrub and short-tree forest habitats from the Río Yaqui in Sonora to the coastal plains of northern Sinaloa. A closely related sister taxon, *C. artus*, replaces *C. goldmani* in riparian communities and more mesic, fully developed short-tree forest. The two species are broadly sympatric for only a 20-km stretch from the vicinity of the Sierra Barabampo into northern Sinaloa. It is evident that the two sister taxa represent more xeric-adapted (*C. goldmani*) and more mesic-adapted (*C. artus*) inhabitants of the coastal plains (Anderson, 1964; Patton, 1969), and that they have met secondarily at the Sierra Barabampo–Río Fuerte filter-barrier. Patton (1969) has described multiple chromosomal races of the northern species (*C. goldmani*), the boundaries of which correspond remarkably with the Río Yaqui, Río Mayo, Río Cuchajaqui, and Río Fuerte. The Sierra Barabampo–Río Fuerte filter-barrier corresponds to older, species-level vicariance between the two species, whereas the Río Fuerte has acted more recently in divergence of chromosomal races within *C. goldmani*.

4.4.7   EXTENT AND SUBDIVISIONS OF THE CHIHUAHUAN DESERT

Our analysis of arid-adapted rodents indicates that the Chihuahuan subregion of the traditional Chihuahuan regional desert (bounded by the Cochise filter-barrier, Río Grande, and Río Conchos) is instead a transitional area between the Sonoran and Chihuahuan deserts and that the core Chihuahuan Desert begins only south of the Río Conchos and extends north along the east side of the Río Grande into Trans-Pecos Texas (Figure 4.8). This remains an hypothesis to be tested with other taxa, particularly plants and reptiles. In such studies, it will be necessary to fully sample all subregions. For example, Walpole et al. (1997) reported genetic distinction (consistent with species-level divergence) between Sonoran and Chihuahuan samples of cactus mice (*Peromyscus eremicus*). They compared samples from southern Arizona with samples from east of the Río Grande in New Mexico and Texas. By not including samples from the Chihuahuan subregion, Walpole et al. (1997) were unable to detect that *P. eremicus* from that subregion were of the Sonoran, not Chihuahuan, form (Riddle et al., 2000a). Had they selected samples of their Chihuahuan Desert *Peromyscus eremicus* from west of the Río Grande to compare with Sonoran Desert samples instead of from east of the Río Grande, they would have concluded that there was no distinction between *P. eremicus* of the two regional deserts. Although they noted that future sampling in southern New Mexico and Chihuahua was necessary to determine the nature of interaction between Sonoran and Chihuahuan forms, they repeatedly stated a genetic dichotomy between Sonoran and Chihuahuan *P. eremicus*, implying that the point of contact was located at the traditional boundary between the two deserts in southwestern New Mexico.

The Cuatro Ciénegas Basin of eastern Coahuila rests astride the Sierra del Carmen–Sierra Madre Oriental filter-barrier, between the Chihuahuan regional desert and the tropical Tamaulipan biotic province. The system of interconnected, travertine-lined pozas (spring-fed pools) and lagunas (spring-fed lakes) of this biodiverse 'biological island' (Contreras-Balderas, 1990) has remained remarkably stable throughout Pleistocene climatic oscillations (Meyer, 1973). At the same time, a combination of archaeological, historical, and current mammalian faunas depicts a dynamic zone of flux between arid and tropical species (Contreras-Balderas et al., 2007).

The Southern Coahuila filter-barrier has long been recognised as a major discontinuity within the Chihuahuan Desert (Baker, 1956; Baker and Greer, 1962; Peterson, 1976; Schmidly, 1977). Hafner et al. (2008) evaluated the systematic status of the yellow-cheeked pocket gopher (*Cratogeomys castanops*) north and south of this filter-barrier, based on morphology, chromosomal complement, and both mtDNA and nuclear DNA sequence data. They concluded that the two represent distinct species (*C. goldmani* to the south of the filter-barrier) and reviewed the distribution of arid-adapted rodents across the filter-barrier. Of the 16 species that occur in the vicinity,

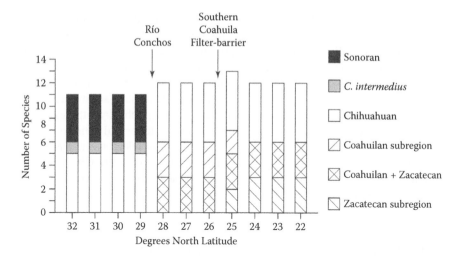

FIGURE 4.10 Shift in proportion of Sonoran (5 spp.) and Chihuahuan (5 spp.) arid-adapted rodents moving from southern New Mexico south through Chihuahua, Coahuila, and Durango to Zacatecas, México; subregions are shown in Figure 4.1. The pocket mouse *Chaetodipus intermedius* occurs in the Sonoran, Chihuahuan, and Trans-Pecos subregions.

six species reach their distributional limits at the filter-barrier and eight species have named subspecies whose distributional boundaries coincide with the filter-barrier (Figure 4.10).

4.5 EVOLUTIONARY PERSPECTIVE ON NORTH AMERICAN WARM DESERTS

It seems appropriate that regional deserts traditionally have been defined based on floral composition, as plants provide the basic habitat supporting regional biotas. However, without phylogeographic studies of wide-ranging plant taxa of the North American deserts, there is a strong possibility that existing regional definitions are biased by phenetic similarities and that homoplasy or conserved plesiomorphic morphology is hiding important evolutionary signals. The discovery of cryptic forms in a variety of animal taxa increases the likelihood that similar homoplasy or conserved morphology exists in desert plants, and it has already been discovered within cryptic forms of Peninsular versus Sonoran desert columnar cacti (*Lophocereus schottii*; Nason et al., 2002). Two methods have been employed to overcome this deficiency in botanical phylogeographic studies: inclusion of a broad array of biological, geological, and climatic data to identify major biotic components (Morrone, 2005); and comparative phylogeographic analysis of a variety of mostly animal taxa (Riddle and Hafner, 2006a). Both approaches have reached the conclusion that there are three major warm regional deserts in North America: the Peninsular, Sonoran, and Chihuahuan regional deserts.

At a finer geographic scale, we have examined phylogeographic patterns amongst a selected subset of arid-adapted rodents in order to evaluate potential filter-barriers between and within neighbouring regional deserts. We posit these filter-barriers as hypotheses deserving of further testing using other taxa. Certainly the most contentious and potentially complicated (temporally and geographically) issues involve the detailed formation of the Baja California Peninsula and Gulf of California, and specifically coalescence of continental land masses transferred across the plate boundary with neovolcanics to form first a peninsular archipelago and eventually the continuous peninsula. The ultimate test of these phylogeographic hypotheses will be geological data supporting or rejecting the hypothesised terrestrial connections and timing of filter-barriers, and fossil evidence supporting or rejecting the past distributional reconstructions. We strongly encourage further investigation by geologists and palaeontologists as regards the timing and detailed geomorphology of perhaps multiple, temporally-nested vicariant events and dispersal corridors associated with these geomorphic events and filter-barriers.

4.5.1 POTENTIAL HAZARDS OF A PURELY PHENETIC APPROACH

The current definition of boundaries and internal filter-barriers within the regional deserts of North America ignores cryptic phylogeographic patterns by considering only similarity of plant species composition, with plant relationships based solely on morphological comparison. This basic framework not only fails to capture the deeper evolutionary relationships amongst and within the regional deserts; it potentially misleads sampling schemes for phylogeographic and ecological studies. For example, the incomplete subregional sampling by Walpole et al. (1997; discussed above) would have resulted in completely different results had the Chihuahuan subregion, rather than the Trans-Pecos subregion, been selected to represent the Chihuahuan Desert, and the implied assumption of the presence of the Chihuahuan form in the Chihuahuan subregion was subsequently proved to be incorrect (Riddle et al., 2000a). Similarly, studies concerning structure and assembly rules of desert mammal communities (e.g., Kelt, 1999; Brown et al., 2000) may have been hindered by strict reliance on the original delineation of regional deserts by Shreve (1942). An implicit assumption of homogeneity within regions and across the Peninsular and Sonoran regions obscured the evolutionary distinctness of the Peninsular Desert (which was not sampled) and led to only a single representative sample of each of the Sonoran (*sensu stricto*) and Chihuahuan Deserts from Mexico (Riddle and Hafner, 1999; Hafner and Riddle, 2005). By sampling only the species-depauperate northern periphery of each of these regions and by relying on previous phenetic studies of mammals that assumed regional homogeneity, these studies failed to include 28 (62%; Brown and Kurzius, 1987) to 24 (53%; Kelt et al., 1996) of the arid-adapted rodents that we list from the southern desert regions (Hafner and Riddle, 2005).

4.5.2 ADVANTAGES OF AN EVOLUTIONARY PERSPECTIVE

Phylogeographic investigation of subsets of the regional desert biota has already resulted in a dramatic shift in our view of the evolution of North American warm deserts. Previously two regions were recognised (Sonoran and Chihuahuan) and the Peninsular flora and fauna were considered to be recently derived from the Sonoran (late Pleistocene). After speciation of Gulf island endemics, a sharp boundary (of debatable location) was thought to divide the Sonoran and Chihuahuan regions and each was regarded as a relatively homogeneous region. The current view includes vicariant events of far deeper history (Miocene and Pliocene age), with multiple ancient origins of the Peninsular biota, very recent (late Pleistocene or Holocene) origin of most island endemics, large areas of transition between the Peninsular and Sonoran, and Sonoran and Chihuahuan regions, and strong substructuring within all three regions. This rich evolutionary history should be reflected in the classification of the North American deserts.

4.5.3 Conservation Implications

If taxa other than mammals also document large zones of transition between neighbouring desert regions and more restricted zones of intermixing at more recent filter-barriers, this information must be included in selection of areas for conservation priority. In the case of arid-adapted rodents, high species diversity signals both centres of areas of endemicity (deserving of conservation priority) and zones of overlap that are peripheral to contributing regions. It would be unwise to apply a simple criterion of high species richness (biodiversity) to these transitional zones as they are by definition the most ephemeral habitats under shifting climates.

4.6 INTEGRATION OF GEOLOGICAL, FOSSIL, AND PHYLOGEOGRAPHIC ANALYSES

Historical biogeography has traditionally depended on a firm foundation in historical geology and fossil data to reconstruct the past distributions of living species. As phylogeographic analyses have increased in sophistication and ability to detect even subtle and previously cryptic biotic shifts, new avenues of integration at the interface of geological, fossil, and molecular phylogeographic studies are opening. Indirect methods of relating fossil and living forms have included comparison of fossil, historical, and current faunal surveys at single sites (e.g., examples of Holocene faunal shifts at the Cochise and Sierra del Carmen–Sierra Madre Oriental filter-barriers described previously; Holmgren et al., 2007; Contreras-Balderas et al., 2007). Hunter et al. (2001) used statistically robust differences in guard cell size in the leaves of living and fossil samples of a widespread desert shrub, creosote bush (*Larrea tridentata*), to infer the changing distribution of ploidy races since the Last Glacial Maximum, and Smith and Betancourt (2006) compared faecal pellet size amongst living and fossil woodrats (or packrats; *Neotoma*) from packrat middens to explore changes in body size and species distributions during the Holocene.

Hadley et al. (2003) demonstrated that a more direct method of linking fossil and molecular phylogeographic data is practical: they extracted ancient DNA from fossil and modern teeth of tuco-tucos (*Ctenomys sociabilis*) from a late Holocene site in Argentina and were able to trace the genetic history of the population through 1 Ka. Chan et al. (2005, 2006) expanded these studies back to 10 Ka and were able to reliably correlate a population genetic bottleneck in the species to the massive eruption of an Andean volcano (Villarosa et al., 2006). Although Reiss (2006) recommends caution in this exciting new field, specifically in the use of material contained in middens, it is likely that studies currently underway by Poinar, Betancourt, Patton, and Smith (F.A. Smith, pers. comm., 2008) at the McMaster Ancient DNA Centre (McMaster University; http://socserv.mcmaster.ca/adna/Research_projects_packrat. htm) will eventually be able to use the extensive collection of American packrat middens, which offers 'the richest archive of dated, identified and well-preserved plant and animal remains in the World' (Pearson and Betancourt, 2002). Unfortunately, there are still relatively few reported collections of packrat middens from Mexico, which contains the majority of North American aridlands.

Secondary-level analyses of molecular phylogeographic patterns, such as the study of mitonuclear discordance at filter-barriers, promise to produce more questions regarding the sources of such discordances that will depend on more precise geological and palaeontological data for full resolution. Such discordant patterns between the nuclear and mitochondrial genomes may be due to incomplete sorting or mitochondrial introgression following a relatively recent vicariant event and secondary contact, or may instead reflect multiple, temporally-nested vicariant events recorded only in the faster-evolving mtDNA genome.

Several lines of geological and palaeontological research in North American aridlands are immediately suggested. In light of the new realisation of increased temporal and geographic complexity of the formation of the Baja California Peninsula and the Gulf of California, a comprehensive and detailed geological summary of this region is needed to identify specific sites requiring field study. Field study should focus on reconstruction of the distribution of marine and terrestrial environments during the last 10–20 Ma, the specific reconstruction of the peninsular archipelago, and the relationships of these archipelago components to the Trans-Mexico Volcanic Belt, adjacent Sonoran, or the California continental areas. Palaeontological fieldwork should target collection and analysis of macrofossils (either from screen-washing of fossiliferous deposits or from packrat middens) from major filter-barriers and peripheral margins within and amongst the regional deserts. In both cases, data can immediately test hypotheses already developed from comparative phylogeographic studies.

SUMMARY

Comparative phylogeography has provided a conceptual bridge between traditional ecological and historical biogeography, and between statistical phylogeography and area-based historical biogeography. By identifying congruent phylogeographic patterns across co-distributed taxa, comparative phylogeography builds a predictive framework that may be continually updated and improved with additional taxa. We offer a comparative phylogeographic analysis of the North American aridlands as a case study to demonstrate the predictive potential of this approach. In particular, we note the specific needs for geological and palaeontological study revealed by this analysis. The regional deserts of North America have traditionally been defined on the basis of plant species composition, which in turn has been determined primarily on phenetic bases. The lack of phylogeographic studies of widespread arid-adapted plant taxa has forced the use of alternative measures to approximate the evolutionary distinctiveness and pattern of relationships amongst desert regions and subregions. We review a growing body of knowledge regarding the geomorphology and evolving climate of the regional deserts and argue for an explicitly evolutionary perspective in redefining the extent of North American deserts and the identity and relationships of three warm regional deserts: Peninsular, Sonoran, and Chihuahuan. We further evaluate potential filter-barriers within each of these regional deserts based on a selected subset of arid-adapted rodents. More geological and palaeontological investigation is needed to evaluate the timing and nature of multiple, temporally-nested vicariant and dispersal events associated with putative filter-barriers between and within neighbouring regional deserts.

ACKNOWLEDGEMENTS

We are grateful to Paul Upchurch for his invitation to participate in this symposium, and to him and an anonymous reviewer for comments on an earlier draft. This research was supported by National Science Foundation grants DEB-0236957 to DJH and DEB-0237166 to BRR.

LITERATURE CITED

Alexander, L.F. and Riddle, B.R. (2005) Phylogenetics of the New World rodent family Heteromyidae. *Journal of Mammalogy*, 86: 366–379.

Anderson, C.A. (1950) The 1940 E. W. Scripps cruise to the Gulf of California. Part I: Geology of the islands and neighboring land areas. *Geological Society of American Memoirs*, 43: 1–53.

Anderson, S. (1964) The systematic status of *Perognathus artus* and *Perognathus goldmani* (Rodentia). *American Museum Novitates*, 2184: 1–27.

Arriaga, L., Aguilara, C., Espinosa-Organista, D., and Jiménez, R. (1997) Regionalización ecológica y biogeográfica de México. *Taller de la Comisión Nacional para el Conocimiento y Uso de la Biodiversidad (Conabio)*, México, DF.

Avise, J.C. (2000) *Phylogeography: The History and Formation of Species*, Harvard University Press, Cambridge, MA.

Avise, J.C., Arnold, J., Ball, R.M., Bermingham, E., Lamb, T., Neigel, J.E., Reeb, C.A., and Saunders, N.C. (1987) Intraspecific phylogeography—the mitochondrial-DNA bridge between population-genetics and systematics. *Annual Review of Ecology and Systematics*, 18: 489–522.

Axelrod, D.I. (1958) Evolution of the Madro-Tertiary geoflora. *Botanical Review*, 24: 433–509.

Axelrod, D.I. (1983) Paleobotanical history of the western deserts. In *Origin and Evolution of Deserts* (eds. S.G. Wells and D.R. Haragan), University of New Mexico Press, Albuquerque, pp. 113–129.

Bailey, R.G. (1997) Ecoregions of North America. *USDA Forest Service, Miscellaneous Publication* 1548.

Baker, R.H. (1956) Mammals of Coahuila, México. *University of Kansas Publications, Museum of Natural History*, 9: 125–335.

Baker, R.H. and Greer, J.K. (1962) Mammals of the Mexican state of Durango. *Publications of the Museum, Michigan State University, Biological Series*, 2: 25–154.

Blair, W.N. (1978) Gulf of California in Lake Mead area of Arizona and Nevada during late Miocene time. *American Association of Petroleum Geologists Bulletin*, 62: 1159–1170.

Boehm, M.C. (1984) An overview of the lithostratigraphy, biostratigraphy, and paleoenvironments of the Late Neogene San Felipe marine sequence, Baja California, Mexico. In *Geology of the Baja California Peninsula* (ed. V.A. Frizzell, Jr.), Society of Economic Paleontology and Mineralogy, 39, pp. 253–265.

Boehm, M.C. (1987) Evidence for a north-verging mid-to-late Miocene proto-Gulf of California. *Geological Society of America, Abstracts with Program*, 19: 594.

Brooks, D.R., van Veller, M.G.P., and McLennan, D.A. (2001) How to do BPA, really. *Journal of Biogeography*, 28: 345–358.

Brown, D.E. (ed.) (1982) Biotic communities of the American Southwest—United States and Mexico. *Desert Plants* 4: 1–4.

Brown, J.H. and Kurzius, M.A. (1987) Composition of desert rodent faunas: Combinations of coexisting species. *Annales Zoologici Fennici*, 24: 227–237.

Brown, J.H. and Lomolino, M.V. (1998) *Biogeography*, 2nd edition. Sinauer, Sunderland, MA.

Brown, J.H., Fox, B.J., and Kelt, D.A. (2000) Assembly rules: Desert rodent communities are structured at scales from local to continental. *American Naturalist*, 156: 314–321.

Brown, J.H., Valone, T.J., and Curtin, C.G. (1997) Reorganization of an arid ecosystem in response to recent climate change. *Proceedings of the National Academy of Sciences, USA*, 94: 9729–9733.

Buising, A.V. (1990) The Bouse Formation and bracketing units, southeastern California and western Arizona: Implications for the evolution of the proto-Gulf of California and the lower Colorado River. *Journal of Geophysical Research*, 95: 111–132.

Campos-Enriquez, J.O., Ortega-Ramírez, J., Alatriste-Vilchis, D., Cruz-Gática, R., and Cabral-Cano, E. (1999) Relationship between extensional tectonic style and the paleoclimatic elements at Laguna El Fresnal, Chihuahua Desert, Mexico. *Geomorphology*, 28: 75–94.

Capra, L., Macías, J.L., Espíndola, J.M., and Siebe, C. (1998) Holocene plinian eruption of La Virgen volcano, Baja California, Mexico. *Journal of Volcanology and Geothermal Research*, 80: 239–266.

Carreño, A.L. and Helenes, J. (2002) Geology and ages of the islands. In *A New Island Biogeography of the Sea of Cortéz* (eds. T.J. Case, M.L. Cody, and E. Ezcurra), Oxford University Press, New York, pp. 14–40.

Casas-Andreu, G. and Reyna-Trujillo, T. (1990) Herpetofauna (anfibios y reptiles). Map IV.8.6. In *Atlas Nacional de México, vol. III*, Instituto de Geografía, Universidad Nacional Autónoma de México, México, DF.

Castoe, T.A., Spencer, C.L., and Parkinson, C.L. (2007) Phylogeographic structure and historical demography of the western diamondback rattlesnake (*Crotalus atrox*): A perspective on North American desert biogeography. *Molecular Phylogenetics and Evolution*, 42: 193–212.

Chamberlin, R.V. (1924) Expedition of the California Academy of Sciences to the Gulf of California in 1921—The spider fauna of the shores and islands of the Gulf of California. *Proceedings of the California Academy of Sciences*, 12: 561–694.

Chan, Y.L., Anderson, C.N.K., and Hadly, E.A. (2006) Bayesian estimation of the timing and severity of a population bottleneck from ancient DNA. *PLoS Genetics*, 2: 451–460.

Chan, Y.L., Lacey, E.A., Pearson, O.P., and Hadly, E.A. (2005) Ancient DNA reveals Holocene loss of genetic diversity in a South American rodent. *Biology Letters*, 1: 423–426.

Cody, M.L. (1983) The land birds. In *Island Biogeography in the Sea of Cortéz* (eds. T.J. Case and M.L. Cody), University of California Press, Berkeley, pp. 210–245.

Cody, M.L. and Velarde, E. (2002) Land birds. In *A New Island Biogeography of the Sea of Cortéz* (eds. T.J. Case, M.L. Cody, and E. Ezcurra), Oxford University Press, New York, pp. 271–312.

Connell, S.D., Hawley, J.W., and Love, D.W. (2005) Late Cenozoic drainage development in the southeastern Basin and Range of New Mexico, southeasternmost Arizona, and western Texas. In *New Mexico's Ice Ages* (eds. S.G. Lucas, G.S. Morgan, and K.E. Zeigler), New Mexico Museum of Natural History and Science Bulletin 28, pp. 125–150.

Contreras-Balderas, A.J., Hafner, D.J., Lopez-Soto, J.H., Torres-Ayala, J.M., and Contreras-Arquieta, S. (2007) Mammals of the Cuatro Ciénegas Basin, Coahuila, Mexico. *Southwestern Naturalist*, 52: 400–409.

Contreras-Balderas, S. (1990) Importancia biota endémica y perspectivas actuales en el Valle de Cuatrociénegas, Coahuila, México. In *Areas Naturales Protegidas en México y Especies en Extinción* (eds. C.R. José and R.A. Fermín), Universidad Nacional Autónoma de México, pp. 15–23.

Crews, S.C. and Hedin, M. (2006) Studies of morphological and molecular phylogenetic divergence in spiders (Araneae: *Homalonychus*) from the American southwest, including divergence along the Baja California peninsula. *Molecular Phylogenetics and Evolution*, 38: 470–487.

Croizat, L. (1952) *Manual of Phytogeography*, Dr. W. Junk, The Hague.

Croizat, L. (1958) *Panbiogeography*. 2 vol., published by the author, Caracas.

Croizat, L. (1960) *Principia Botanica*, published by the author, Caracas.

Croizat, L. (1964) *Space, Time, Form: The Biological Synthesis*, published by the author, Caracas.

Devitt, T.J. (2006) Phylogeography of the Western Lyre snake (*Trimorphodon biscutatus*): testing aridland biogeographical hypotheses across the Nearctic-Neotropical transition. *Molecular Ecology*, 15: 4387–407.

Dickey, K.J. (1983) *A Natural History Guide to Baja California*, published by the author, Chula Vista, CA.

Dunbier, R. (1968) *The Sonoran Desert: Its Geography, Economy, and People*. University of Arizona Press, Tucson.

Durham, J.W. and Allison, E.C. (1960) Geological history of Baja California and its marine environments. Symposium: The biogeography of Baja California and adjacent seas. *Systematic Zoology*, 9: 47–91.

Eberly, L.B. and Stanley, T.B., Jr. (1978) Cenozoic stratigraphy and geologic history of southwestern Arizona. *Geological Society of American Bulletin*, 89: 921–940.

Faulds, J.E., Wallace, M.A., Gonzalez, L.A., and Heizler, M.T. (2001) Depositional environment and paleogeographic implications of the late Miocene Hualapai Limestone, northwest Arizona and southern Nevada. In *The Colorado River: Origin and Evolution: Grand Canyon, Arizona* (eds. R.A. Young and E.E. Spamer), Grand Canyon Association Monograph 12, pp. 8–87.

Fehlberg, S.D. and Ranker, T.A. (2007) Phylogeny and biogeography of *Encelia* (Asteraceae) in the Sonoran and Peninsular Deserts based on multiple DNA sequences. *Systematic Botany*, 32: 692–699.

Ferrusquía-Villafranca, I. (1990) Regionalización biogeográfica. Map IV.8.10. In *Atlas Nacional de México, vol. III*, Instituto de Geografía, Universidad Nacional Autónoma de México, México, D.F.

Ferrusquía-Villafranca, I., González Guzmán, L.I., and Cartron, J.-L.E. (2005) Northern Mexico's landscape, part I: The physical setting and constraints on modeling biotic evolution. In *Biodiversity, Ecosystems, and Conservation in Northern Mexico* (eds. J.-L.E. Cartron, G. Ceballos, and R.S. Felger), Oxford University Press, New York, pp. 11–38.

Findley, J.S. (1969) Biogeography of southwestern boreal and desert mammals. In *Contributions in Mammalogy* (ed. J.K. Jones), University of Kansas Miscellaneous Publications of the Museum of Natural History, 51, pp. 113–128.

Flynn, J.J. and Novacek, M.J. (1984) Early Eocene vertebrates from Baja California: Evidence for intracontinental age correlations. *Science*, 224: 151–153.

Gantenbein, B., Fet, V. and Barker, M.D. (2001) Mitochondrial DNA reveals a deep, divergent phylogeny in *Centruroides exilicauda* (Wood, 1863) (Scorpiones: Buthidae). In *Scorpions 2002. In Memoriam Gary A. Polis* (eds. V. Fet and P.A. Selden), British Arachnological Society, Burnham Beeches, UK, pp. 235–244.

Garcia, E. and de Byres, Z. F. (1984) *Nuevo Porrua de la Republica Mexicana*. [Climatic map], Editorial Porrua, S. A., Mexico, DF, p. 113.

Garduño-Monroy, H., Vargas-Ledezma, H., and Campos-Enriquez, J.O. (1993) Preliminary geologic studies of Sierra El Aguajito (Baja California, Mexico): A resurgent-type caldera. *Journal of Volcanology and Geothermal Research*, 59: 47–58.

Gastil, G., Minch, J., and Phillips, R.P. (1983) The geology and ages of the islands. In *Island Biogeography in the Sea of Cortéz* (eds. T.J. Case and M.L. Cody), University of California Press, Berkeley, pp. 13–25.

Gastil, J., Neuhaus, J., Cassidy, M., Smith, J., Ingle, J., and Krummenacker, D. (1999) Geology and paleontology of Southwestern Isla Tiburón, Mexico. *Revista Mexicana de Ciencias Geológicas*, 16: 1–34.

Grismer, L.L. (1994) The origin and evolution of the Peninsular herpetofauna of Baja California, México. *Herpetological Natural History*, 2: 51–106.

Grismer, L.L. (2002) A re-evaluation of the evidence for a mid-Pleistocene mid-peninsular seaway in Baja California: A reply to Riddle et al. *Herpetological Review*, 33: 15–16.

Hadley, E.A., van Tuinen, M., Chan, Y., and Heiman, K. (2003) Ancient DNA evidence of prolonged population persistence with negligible genetic diversity in an endemic tuco-tuco (*Ctenomys sociabilis*). *Journal of Mammalogy*, 84: 403–417.

Hafner, D.J. and Riddle, B.R. (1997) Biogeography of Baja California Peninsular Desert mammals. In *Life Among the Muses: Papers in Honor of James S. Findley* (eds. T.L. Yates, W.L. Gannon, and D.E. Wilson), Special Publication, Museum of Southwestern Biology, University of New Mexico 3, pp. 39–65.

Hafner, D.J. and Riddle, B.R. (2005) Mammalian phylogeography and evolutionary history of northern Mexico's deserts. In *Biodiversity, Ecosystems, and Conservation in Northern Mexico* (eds. J.-L.E. Cartron, G. Ceballos, and R.S. Felger), Oxford University Press, New York, pp. 225–245.

Hafner, D.J., Hafner, M.S., Hasty, G.L., Spradling, T.A., and Demastes, J.W. (2008) Evolutionary relationship of pocket gophers (*Cratogeomys castanops* species group) of the Mexican Altiplano. *Journal of Mammalogy*, 89: 190–208.

Hafner, J.C., Light, J.A., Hafner, D.J., Hafner, M.S., Reddington, E., Rogers, D.S., and Riddle, B.R. (2007) Basal clades and molecular systematics of heteromyid rodents. *Journal of Mammalogy*, 88: 1129–1145.

Hall, E.R. (1981) *The Mammals of North America. Volumes 1 and 2*. John Wiley and Sons, New York.

Hawley, J.W., Hibbs, B.J., Kennedy, J.F., Creel, B.J., Remmenga, M.D., Johnson, M., Lee, M.M., and Dinterman, P. (2000) Trans-international boundary aquifers in southwest New Mexico. Technical completion report prepared for U.S. Environmental Protection Agency—Region 6 and the International Boundary and Water Commission.

Helenes, J. and Carreño, A.L. (1999) Neogene sedimentary evolution of Baja California in relation to regional tectonics. *Journal of South American Earth Sciences*, 12: 589–605.

Hennig, W. (1950) *Grundzüge einer Theorie der Phylogenetischen Systematik*. Deutscher Zentralverlag, Berlin.

Hennig, W. (1966) *Phylogenetic Systematics*, 3rd edition (trans. D.D. Davis and R. Zanderl), University of Illinois Press, Urbana.

Hess, H.H. (1962) History of ocean basins. In *Petrological Studies: A Volume in Honor of A. F. Buddington*. (eds. A.E.J. Engel, H.L. James, and B.F. Leonard), Geological Society of America, New York, pp. 599–620.

Holmgren, C.A., Norris, J., and Betancourt, J.L. (2007) Inferences about winter temperatures and summer rains from the late Quaternary record of C_4 perennial grasses and C_3 desert shrubs in the northern Chihuahuan Desert. *Journal of Quaternary Science*, 22: 141–161.

Holt, J.W., Holt, E.W., and Stock, J.M. (2000) An age constraint on Gulf of California rifting from the Santa Rosalía basin, Baja California Sur, Mexico. *Geological Society of America Bulletin*, 112: 540–549.

Hubbard, J.P. (1974) Avian evolution in the aridlands of North America. *Living Bird*, 12: 155–196.

Humphries, C.J. and Parenti, L.R. (1999) *Cladistic Biogeography*, Oxford University Press, Oxford.

Hunt, C.B. (1983) Physiographic overview of our arid lands in the western U.S. In *Origin and Evolution of Deserts* (eds. S.G. Wells and D.R. Haragan), University of New Mexico Press, Albuquerque.

Hunter, K.L., Betancourt, J.L., Riddle, B.R., Van Devender, T.R., Cole, K.L., and Spaulding, W.G. (2001) Ploidy race distributions since the Last Glacial Maximum in the North American desert shrub, *Larrea tridentata*. *Global Ecology and Biogeography*, 10: 521–533.

Hurtado, L.A., Erez, T., Castrezana, S., and Markow, T.A. (2004) Contrasting population genetic patterns and evolutionary histories among sympatric Sonoran Desert cactophilic *Drosophila*. *Molecular Ecology*, 13: 1365–1375.

Ingle, J.C., Jr. (1974) Paleobathymetric history of Neogene marine sediments, Northern Gulf of California. In *The Geology of Peninsular California*, Guidebook 49 of the Pacific Section (eds. G. Gastil and J. Lillegraven), American Association of Petroleum Geologists and Society of Economic Paleontologists and Mineralogists, pp. 121–138.

Ingle, J.C., Jr. (1987) Paleoceanographic evolution of the Gulf of California: Foraminiferal and lithofacies evidence. *Geological Society of America—Abstracts with Programs*, 19: 721.

Instituto de Geografía (1970) *Carta de Climas de la Republica*, Universidad Nacional Autónoma de México, Comision de Estudios del Territorio Nacional, México, DF.

Ives, R.L. (1962) Dating of the 1746 eruption of Tres Vírgenes Volcano, Baja California del Sur, Mexico. *Geological Society of America Bulletin*, 73: 647–648.

Kelt, D.A. (1999) On the relative importance of history and ecology in structuring communities of desert small animals. *Ecography*, 22: 123–137.

Kelt, D.A., Brown, J.H., Heske, E.J., Marquet, P.A., Morton, S.R., Reid, J.R., Rogouin, K.A., and Shenbrot, G. (1996) Community structure of desert small mammals: Comparisons across four continents. *Ecology*, 77: 746–761.

Leaché, A.D., Crews, S.C., and Hickerson, M.J. (2007) Two waves of diversification in mammals and reptiles of Baja California revealed by hierarchical Bayesian analysis. *Biology Letters*, 3: 646–650.

Ledesma-Vázquez, J. (2002) A gap in the Pliocene invasion of seawater to the Gulf of California. *Revista Mexicana de Ciencias Geológicas*, 19: 145–151.

Lindell, J., Méndez-de la Curz, F.R., and Murphy, R.W. (2005) Deep genealogical history without population differentiation: Discordance between mtDNA and allozyme divergence in the zebra-tailed lizard (*Callisaurus draconoides*). *Molecular Phylogenetics and Evolution*, 36: 682–694.

Lindell, J., Ngo, A., and Murphy, R.W. (2006) Deep genealogies and the mid-peninsular seaway of Baja California. *Journal of Biogeography*, 33: 1327–1331.

Lomolino, M.V., Riddle, B.R., and Brown, J.H. (2005) *Biogeography*, 3rd edition, Sinauer, Sunderland, MA.

Lonsdale, P. (1989) Geology and tectonic history of the Gulf of California. In *The Eastern Pacific Ocean and Hawaii* (eds. E.L. Winterer, D.M. Hussong, and R.W. Decker), Geological Society of America North American Geology Vol. N, pp. 499–521.

Lucchitta, I. (1979) Late Cenozoic uplift of the southwestern Colorado Plateau and adjacent lower Colorado River region. *Tectonophysics*, 61: 63–95.

Mack, G.H., Salyards, S.L., McIntosh, W.C., and Leeder, M.R. (1998) Reversal magnetostratigraphy and radioisotopic geochronology of the Plio-Pleistocene Camp Rice and Palomas formations, southern Rio Grande rift. *New Mexico Geological Society Guidebook*, 49: 229–236.

MacMahon, J.A. (1979) North American deserts: Their floral and faunal components. In *Arid-land Ecosystems: Structure, Functioning, and Management* (eds. R.A. Pern and D.W. Goodall), Cambridge University International Biological Programme, 16: 21–82.

Marshall, C.J. and Liebherr, J.K. (2000) Cladistic biogeography of the Mexican transition zone. *Journal of Biogeography*, 27: 203–216.

Mattick, R.E., Olmsted, F.H., and Zohdy, A.A.R. (1973) Geophysical studies in the Yuma area, Arizona and California. *United States Geological Survey Professional Paper*, 726D: 1–36.

McCloy, C. (1984) Stratigraphy and depositional history of the San Jose del Cabo trough, Baja California Sur, Mexico. In *Geology of the Baja California Peninsula* (ed. V.A. Frizzell, Jr.), Society of Economic Paleontology and Mineralogy, vol. 39, pp. 267–273.

M'Closkey, R.T. (1983) *Perognathus baileyi* and jojoba (*Simmondsia chinensis*): A test of their association. *Journal of Mammalogy*, 64: 499–501.

Metzger, D.G. (1968) The Bouse Formation (Pliocene) of the Parker-Blythe-Cibola area, Arizona and California. *United States Geological Survey, Professional Papers*, 600–D: D126–D136.

Meyer, E.R. (1973) Late-Quaternary paleoecology of the Cuatro Cienegas Basin, Coahuila, Mexico. *Ecology*, 54: 982–995.

Miller, W.E. (1977) Pleistocene terrestrial vertebrates from southern Baja California. *Abstracts with Program, Geological Society of America*, 9: 468.

Miller, W.E. (1980) The late Pliocene Las Tunas local fauna from southernmost Baja California, Mexico. *Journal of Paleontology*, 54: 762–805.

Minch, J. and Leslie, T. (1991) *The Baja Highway: A Geology and Biology Field Guide for the Baja Traveler*, John Minch, San Juan Capistrano, CA.

Morafka, D.J. (1974) A biogeographical analysis of the Chihuahuan Desert through its herpetofauna. Unpublished Ph.D. dissertation, University of Southern California.

Morafka, D.J. (1977a) A biogeographical analysis of the Chihuahuan desert through its herpetofauna. *Biogeographica* 9: 1–313.

Morafka, D.J. (1977b) Is there a Chihuahuan Desert? A quantitative evaluation through a herpetofaunal perspective. In *Transactions of the Symposium on the Biological Resources of the Chihuahuan Desert Region, United States and Mexico* (eds. R.H. Wauer and D.H. Riskind), National Park Service Transactions and Proceedings Series 3, Washington, DC, pp. 437–454.

Moritz, C., Lavery, S., and Slade, R. (1995) Using allele frequency and phylogeny to define units for conservation and management. *American Fisheries Society Symposium*, 17: 249–262.

Morrison, R.B. (1965) Quaternary surfaces and associated deposits in Duncan Valley, Arizona-New Mexico. *International Association for Quaternary Research (INQUA), VII Congress, 1965, Guidebook, Field Conference H (Southwestern Arid Lands)*, pp. 66–69.

Morrison, R.B. (1985) Pliocene/Quaternary geology, geomorphology, and tectonics of Arizona. *Geological Society of America Special Paper*, 203: 123–146.

Morrison, R.B. (1991) Quaternary geology of the southern Basin and Range province. In *Quaternary Non-glacial Geology* (ed. R.B. Morrison), The Geology of North America, Geological Society of America, Boulder, CO, K-2: 353–371.

Morrone, J.J. (1994) On the identification of areas of endemism. *Systematic Biology*, 43: 438–441.

Morrone, J.J. (2001a) Biogeografía de América Latina y el Caribe. Zaragoza (España): Manuales y Tesis SEA 3, 148 pp.

Morrone, J.J. (2001b) Toward a cladistic model for the Caribbean subregion: Delimitation of areas of endemism. *Caldasia*, 23: 43–76.

Morrone, J.J. (2004) Panbiogeografía, componentes bióticos y zonas de transición. *Coleopterists Bulletin*, 48: 149–162.

Morrone, J.J. (2005) Hacia una síntesis biogeográfica de México (toward a synthesis of Mexican biogeography). *Revista Mexicana de Biodiversidad*, 76: 207–252.

Morrone, J.J. and Crisci, J.V. (1995) Historical biogeography — Introduction to methods. *Annual Review of Ecology and Systematics*, 26: 373–401.

Morrone, J.J. and Márquez, J. (2001) Halffter Mexican Transition Zone, generalised tracks, and geographical homology. *Journal of Biogeography*, 28: 635–650.

Mullis, K.B, Faloona, F.A., Scharf, S., Saiki, R.K., Horn, G., and Erlich, H.A. (1986) Specific enzymatic amplification of DNA in vitro: The polymerase chain reaction. *Cold Spring Harbor Symposia on Quantitative Biology 1986*.

Murphy, R.W. (1983) Paleobiogeography and patterns of genetic differentiation of the Baja California herpetofauna. *Occasional Papers, California Academy of Sciences*, 137: 1–48.

Murphy, R.W. and Aguirre-León, G. (2002) The nonavian reptiles: Origins and evolution. In *A New Island Biogeography of the Sea of Cortéz* (eds. T.J. Case, M.L. Cody, and E. Ezcurra), Oxford University Press, New York, pp. 181–220.

Nason, J.D., Hamrick, J.L., and Fleming, T.H. (2002) Historical vicariance and postglacial colonization effects on the evolution of genetic structure in *Lophocereus*, a Sonoran Desert columnar cactus. *Evolution*, 56: 2214–2226.

Norris, R.M. and Webb, R.W. (1976) *Geology of California*, John Wiley and Sons, New York.

Olmsted, F.H., Loeltz, O.J., and Irelan, B. (1973) Geohydrology of the Yuma Area, Arizona and California. *United States Geological Survey Professional Paper*, 486-H: 1–227.

Orr, R.T. (1960) An analysis of the recent land mammals. Symposium: The biogeography of Baja California and adjacent seas. *Systematic Zoology*, 9: 171–179.

Ortega-Gutiérrez, F. and Guerrero-Garcia, J.C. (1982) The geologic regions of Mexico. In *Perspectives in Regional Geological Synthesis* (ed. A.R. Palmer), D-NAG Special Publication 1, Geological Society of America, Boulder, CO, pp. 99–104.

Ortlieb, L. (1991) Quaternary vertical movements along the coasts of Baja California and Sonora. In *The Gulf and Peninsular Province of the Californias* (eds. J.P. Dauphin and B.R.T. Simoneit), American Association of Petroleum Geologists, Memoir 47, pp. 447–480.

Patton, J.L. (1969) Chromosome evolution in the pocket mouse, *Perognathus golmani* Osgood. *Evolution*, 23: 645–662.

Pazzaglia, F.J. and Hawley, J.S. (2004). Neogene (rift flank) and Quaternary geology and geomorphology. In *The Geology of New Mexico: A Geologic History* (eds. G.H. Mack and K.J. Giles), New Mexico Geological Society, Special Publication 11, Socorro, NM, pp. 407–438.

Pearson, S. and Betancourt, J.L. (2002) Understanding arid environments using fossil rodent middens. *Journal of Arid Environments*, 50: 499–511.

Petersen, M.D. and Wesnousky, S.G. (1994) Fault slip rates and earthquake histories for active faults in southern California. *Bulletin of the Seismological Society of America*, 84: 1608–1649.

Peterson, M.K. (1976) The Rio Nazas as a factor in mammalian distribution in Durango, Mexico. *The Southwestern Naturalist*, 20: 495–502.

Platnick, N.I. and Nelson, G. (1978) A method of analysis for historical biogeography. *Systematic Zoology*, 27: 1–16.

Prothero, D.R. (1998) The chronological, climatic, and paleogeographic background to North American mammalian evolution. In *Evolution of Tertiary Mammals of North America* (eds. C.M. Janis, K.M. Scott, and L.L. Jacobs), Cambridge University Press, Cambridge, MA, pp. 9–36.

Ramírez-Pulido, J. and Castro-Campillo, A. (1990) Regionalización mastofaunística (mamíferos). In *Atlas Nacional de México*, vol. III, Instituto de Geografía, Universidad Nacional Autónoma de México, México, DF, Map IV.8.8.A.

Reiss, R.A. (2006) Ancient DNA from ice age insects: Proceed with caution. *Quaternary Science Reviews*, 25: 1877–1893.

Rhode, D. (2002) Early Holocene juniper woodland and chaparral taxa in the central Baja California peninsula, Mexico. *Quaternary Research*, 57: 102–108.

Riddle, B.R. and Hafner, D.J. (1999) Species as units of analysis in ecology and biogeography: Time to take the blinders off. *Global Ecology and Biogeography*, 8: 433–441.

Riddle, B.R. and Hafner, D.J. (2004) The past and future roles of phylogeography in historical biogeography. In *Frontiers of Biogeography* (eds. M.V. Lomolino and L.R. Heaney), Sinauer, Sunderland, MA, pp. 93–110.

Riddle, B.R. and Hafner, D.J. (2006a) A step-wise approach to integrating phylogeographic and phylogenetic biogeographic perspectives on the history of a core North American warm deserts biota. *Journal of Arid Environments*, 66: 435–461.

Riddle, B.R. and Hafner, D.J. (2006b) Biogeografía y biodiversidad de los desiertos cálidos del norte de México y soroeste de Estados Unidos. *New Mexico Museum of Natural History and Science Bulletin* 32: 57–65.

Riddle, B.R. and Hafner, D.J. (2007) Phylogeography in historical biogeography: Investigating the biogeographic histories of populations, species, and young biotas. In *Biogeography in a Changing World* (eds. M.C. Ebach and R.S. Tangney), The Systematics Association Special Volume Series no. 70, CRC Press, Boca Raton, FL, pp. 161–176.

Riddle, B.R., Hafner, D.J., and Alexander, L.F. (2000a) Phylogeography and systematics of the *Peromyscus eremicus* species group and the historical biogeography of North American warm regional deserts. *Molecular Phylogenetics and Evolution*, 17: 145–160.

Riddle, B.R., Hafner, D.J., and Alexander, L.F. (2000b) Comparative phylogeography of Bailey's pocket mouse (*Chaetodipus baileyi*) and the *Peromyscus eremicus* species group: historical vicariance of the Baja California Peninsular Desert. *Molecular Phylogenetics and Evolution*, 17: 161–172.

Riddle, B.R., Hafner, D.J., Alexander, L.F., and Jaeger, J.R. (2000c) Cryptic vicariance in the historical assembly of a Baja California Peninsular Desert biota. *Proceedings of the National Academy of Sciences of the United States of America*, 97: 14438–14443.

Riginos, C. (2005) Cryptic vicariance in Gulf of California fishes parallels vicariance patterns found in Baja California mammals and reptiles. *Evolution*, 59: 2678–2690.

Roberts, N.C. (1989) *Baja California Plant Field Guide*. Natural History, La Jolla, CA.

Rodríguez-Robles, J.A. and De Jesús-Escobar, J.M. (2000) Molecular systematics of New World gopher, bull, and pinesnake (*Pituophis*: Colubridae), a transcontinental species complex. *Molecular Phylogenetics and Evolution*, 14: 35–50.

Rosen, D.E. (1978) Vicariant patterns and historical explanation in biogeography. *Systematic Zoology*, 27: 159–188.

Rosen, B.R. (1988) From fossils to earth history: Applied historical biogeography. In *Analytical Biogeography: An Integrated Approach to the Study of Animal and Plant Distributions* (eds. A.A. Meyers and P.S. Giller), Chapman and Hall, London, pp. 437–481.

Rzedowski, J. (1973) Geographical relationships of the flora of Mexican dry regions. In *Vegetation and Vegetational History of Northern Latin America* (ed. A. Graham), Elsevier, Amsterdam, pp. 61–72.

Rzedowski, J. (1978) *La Vegetación de México*, Limusa-Wiley, México, DF.

Rzedowski, J. and Reyna-Trujillo, T. (1990) Tópicos biogeográficos. In *Atlas Nacional de México*, vol. III, Instituto de Geografía, Universidad Nacional Autónoma de México, México, DF, Map IV.8.3.

Savage, J.M. (1960) Evolution of a Peninsular herpetofauna. Symposium: The biogeography of Baja California and adjacent seas. *Systematic Zoology*, 9: 184–212.

Sawlan, M.G. (1986) Petrogenesis of Late Cenozoic volcanic rocks from Baja California Sur, Mexico. University of California, Santa Cruz: Unpublished PhD Thesis, 174 pp.

Schmidly, D.J. (1977) Factors governing the distribution of mammals in the Chihuahuan Desert region. In *Transactions of the Symposium on the Biological Resources of the Chihuahuan Desert Region, United States and Mexico* (eds. R.H. Wauer and D.H. Riskind), National Park Service Transactions and Proceedings Series 3, Washington, DC, pp. 163–192.

Schmidly, D.J., Wilkins, K.T., and Derr, J.N. (1993) Biogeography. In *Biology of the Heteromyidae* (eds. H.H. Genoways and J.H. Brown), Special Publication, The American Society of Mammalogists 10, Provo, UT, pp. 319–356.

Schmidt, E.K. (1975) Plate tectonics, volcanic petrology, and ore formation in the Santa Rosalía area, Baja California peninsula. *Society of Economic Paleontologists and Mineralogists, Special Publication* 39: 237–251.

Schmidt, R.H., Jr. (1989) The arid zones of Mexico: Climatic extremes and conceptualization of the Sonoran Desert. *Journal of Arid Environments*, 16: 241–256.

Schmitt, A.K., Stockli, D.F., and Hausback, B.P. (2006) Eruption and magma crystallization ages of Las Tres Vírgenes (Baja California) constrained by combined $^{230}Th/^{238}U$ and (U-Th)/He dating of zircon. *Journal of Volcanology and Geothermal Research*, 158: 281–295.

Sherbrooke, W.C. (1976) Differential acceptance of toxic jojoba seed (*Simmondsia chinensis*) by four Sonoran Desert heteromyid rodents. *Ecology*, 57: 596–602.

Shreve, F. (1942) The desert vegetation of North America. *Botanical Review*, 8: 195–246.

Shreve, F. (1951) *Vegetation of the Sonoran Desert*, Carnegie Institution of Washington, Washington, DC, Publication no. 591.

Shreve, F. and Wiggins, I.L. (1964) *Vegetation and Flora of the Sonoran Desert*, Stanford University Press, Stanford, CA.

Sinclair, E.A., Bezy, R.L., Bolles, K., Camarillo R.J.L., Crandall, K.A., and Sites, J.W., Jr. (2004) Testing species boundaries in an ancient species complex with deep phylogeographic history: Genus *Xantusia* (Squamata: Xantusiidae). *The American Naturalist*, 164: 396–414.

Sissom, W.D. and Hendrixson, B.E. (2005) Scorpion biodiversity and patterns of endemism in northern Mexico. In *Biodiversity, Ecosystems, and Conservation in Northern Mexico* (eds. J.-L.E. Cartron, G. Ceballos, and R.S. Felger), Oxford University Press, New York, pp.122–137.

Smith, F.A. and Betancourt, J.L. (2006) Predicting woodrat (*Neotoma*) response to anthropogenic warming from studies of the palaeomidden record. *Journal of Biogeography*, 33: 2061–2076.

Smith, G.I. and Street-Perrott, F.A. (1983) Pluvial lakes of the western United States. In *Late-Quaternary Environments of the United States*, vol. 1 (ed. H.E. Wright, Jr.), Longman Group, London.

Smith, H.M. (1941) Las provincias bióticas de México, según la distribución geográfica de ls lagartijas del género *Sceloporus*. *Anales de la Escuela Nacional de Ciencias Biológicas*, 2: 103–110.

Smith, J.T. (1984) Miocene and Pliocene marine mollusks and preliminary correlations, Vizcaino Peninsula to Arroyo la Purisima, northwestern Baja California Sur, Mexico. In *Geology of the Baja California Peninsula* (ed. V.A. Frizzell, Jr.), Pacific Section, Society of Economy Paleontologists and Mineralogists, Los Angeles, pp. 197–217.

Smith, J.T. (1991) Cenozoic marine mollusks and paleogeography of the Gulf of California. In *The Gulf and Peninsular Province of the Californias* (eds. J.P. Dauphin and B.R.T. Simoneit), American Association of Petroleum Geologists, Tulsa, OK, Memoir 47, pp. 637–666.

Spencer, J.E. and Normark, W.R. (1989) Neogene plate-tectonic evolution of the Baja California Sur continental margin and the southern Gulf of California, Mexico. In *The Eastern Pacific Ocean and Hawaii* (eds. E.L. Winterer, D.M. Hussong, and R.W. Decker), The Geology of North America N, Geological Society of America, Boulder, CO, pp. 489–498.

Spencer, J.E. and Pearthree, P.A. (2001) Headward erosion versus closed-basin spillover as alternative causes of Neogene capture of the ancestral Colorado River by the Gulf of California. In *The Colorado River: Origin and Evolution: Grand Canyon, Arizona* (eds. R.A. Young and E.E. Spamer), Grand Canyon Association Monograph 12, Grand Canyon, AZ, pp. 215–219.

Spencer, J.E. and Pearthree, P.A. (2005) Abrupt initiation of the Colorado River and initial incision of the Grand Canyon. *Arizona Geology*, 35: 1–4.

Stock, J.M. and Hodges, K.V. (1989) Pre-Pliocene extension around the Gulf of California and the transfer of Baja California to the Pacific Plate. *Tectonics*, 8: 99–115.

Strain, W.S. (1966) Blancan mammalian fauna and Pleistocene formations, Hudspeth County, Texas. *Bulletin of the Texas Memorial Museum*, 10: 1–55.

Strain, W.S. (1971) Late Cenozoic bolson integration in the Chihuahua tectonic belt. *West Texas Geological Society*, 71–59: 167–173.

Swanson, E.R. and McDowell, F.W. (1984) Calderas of the Sierra Madre Occidental volcanic field, western Mexico. *Journal of Geophysical Research*, 89: 8787–8799.

Thompson, R.W. (1968) Tidal flat sedimentation on the Colorado River Delta, northwestern Gulf of California. *Geological Society of America Memoirs*, 10: 1–133.

Thornwaite, C.W. (1931) Climates of North America. *Geographical Review*, 21: Plate III.

Truxal, F.S. (1960) The entomofauna with special reference to its origin and affinities. Symposium: The biogeography of Baja California and adjacent seas. *Systematic Zoology*, 9: 165–170.

Upton, D.E. and Murphy, R.W. (1997) Phylogeny of the side-blotched lizards (Phrynosomatidae: *Uta*) based on mtDNA sequences: Support for a midpeninsular seaway in Baja California. *Molecular Phylogenetics and Evolution*, 8: 104–113.

Van de Kamp, P.C. (1973) Holocene continental sedimentation in the Salton Basin, California: A reconnaissance. *Geological Society of America Bulletin*, 84: 727–848.

Villarosa, G., Outes, V., Hajduk, A., Montero, E.C., Sellés, D., and Crivelli, M.F.E. (2006) Explosive volcanism during the Holocene in the Upper Limay River Basin: The effects of ashfalls on human societies, Northern Patagonia, Argentina. *Quaternary International*, 158: 44–57.

Walpole, D.K., Davis, S.K., and Greenbaum, I.F. (1997) Variation in mitochondrial DNA in populations of *Peromyscus eremicus* from the Chihuahuan and Sonoran deserts. *Journal of Mammalogy*, 78: 397–404.

Wauer, R.H. and Riskind, D.H. (eds.) (1977) *Transactions of the Symposium on the Biological Resources of the Chihuahuan Desert Region, United States and Mexico*, National Park Service Transactions and Proceedings Series 3, Washington, DC.

Webb, T., III and Bartlein, P.J. (1992) Global changes during the last 3 million years: Climatic controls and biotic responses. *Annual Review of Ecology and Systematics*, 23: 141–173.

Wilkins, D.E. and Currey, D.R. (1997) Timing and extent of late Quaternary paleolakes in the Trans-Pecos Closed Basin, west Texas and south-central New Mexico. *Quaternary Research*, 47: 306–315.

Williams, S.C. (1980) Scorpions of Baja California, Mexico, and adjacent islands. *Occasional Papers of the California Academy of Sciences*, 135: 1–127.

Wood, D.A., Fishers, R.N., and Reeder, T.W. (2008) Novel patterns of historical isolation, dispersal, and secondary contact across Baja California in the Rosy Boa (*Lichanura trivirgata*). *Molecular Phylogenetics and Evolution*, 46: 484–502.

Woodburne, M.O. (2004) Global events and the North American mammalian biochronology. In *Late Cretaceous and Cenozoic Mammals of North America: Biostratigraphy and Geochronology* (ed. M.O. Woodburne), Columbia University Press, New York, pp. 315–343.

Zink, R.M., Barrowclough, G.F., Atwood, J.L., and Blackwell-Rago, R.C. (2000) Genetics, taxonomy, and conservation of the threatened California Gnatcatcher. *Conservation Biology*, 14: 1394–1405.

5 Integrating GIS and Phylogenetic Biogeography to Assess Species-Level Biogeographic Patterns
A Case Study of Late Devonian Faunal Dynamics

Alycia L. Stigall

CONTENTS

5.1   INTRODUCTION

The fossil record contains a rich history of shifting geographic ranges of species in the ancient past (Lieberman, 2003). Quantifying the geographic ranges of species in the fossil record is currently an underdeveloped yet promising area of study. Palaeobiogeographic studies have typically examined shifts in geographic range over large timescales (stages or periods), mainly of higher taxa (Boucot, 1975). Range reconstruction methods using Geographic Information Systems (GIS), however, provide promising new opportunities to quantify ranges of individual fossil species in addition to higher taxa across temporal intervals approximating that of biostratigraphic zones (Rode and Lieberman, 2004, 2005; Stigall Rode and Lieberman, 2005a, 2005b). Recent use of GIS in palaeontology, including reconstructing the species ranges of Palaeozoic invertebrates through multiple temporal intervals (Rode and Lieberman, 2000, 2004, 2005), has begun to surpass cataloguing species occurrences and to produce data to test palaeoecological and palaeobiogeographic hypotheses. Palaeobiogeographic ranges of higher taxa, such as families and orders, contribute important information about the evolutionary history of clades and their co-evolution with the Earth.

Characterisation of the ranges of individual fossil species, however, provides additional insight into evolutionary processes such as speciation and extinction. Because species are the entities through which the macroevolutionary phenomena of speciation and extinction occur, quantification of species ranges provides key information for assessing evolutionary patterns. In particular, quantifying changes in the spatial distribution of species ranges through time can provide insight into biogeographic feedback onto palaeoecology and macroevolution. The geographic range of a fossil species, as for a modern species, represents the geographic limits of the realised niche of that taxon (Brooks and McLennan, 1991, 2002; Lomolino et al., 2006). Quantifying the spatial shifts in these distributions over geologic time, therefore, can provide information about changing environmental tolerances or (more commonly) habitat tracking of a stable species through time (Stigall Rode and Lieberman, 2005b). Studying ranges of individual species, therefore, provides direct information on the interplay between biogeographic and palaeoecological processes.

Furthermore, species ranges when combined with species-level phylogenetic information can provide insight into the relationship amongst the macroevolutionary processes of speciation, extinction, and biogeography. A robust species-level phylogenetic hypothesis will provide a rigorous evolutionary framework in which to interpret specific shifts in geographic ranges of closely related species (Wiley and Mayden, 1985). In addition, the use of species-level phylogenies within a phylogenetic biogeographic framework further provides analytical data on the interaction between biogeography and evolution. In particular, style of speciation, episodes of vicariance or dispersal, and the role of tectonic or environmental drivers of biodiversity change during intervals of biodiversity crisis may be quantified (Lieberman and Eldredge, 1996; Rode and Lieberman, 2005; Stigall Rode and Lieberman, 2005a).

Quantifying the interaction between species range expansion and contraction and macroevolution in the fossil record has the potential to provide insight into the long-term effects of range changes, such as expansion and contraction, during the modern biodiversity crisis. The fossil record provides a rich history of range expansions and interbasinal species invasions that can be studied as analogues of modern events to characterise the long-term effects of modern ecosystem threats such as invasive species. Monitoring and understanding the reasons for geographic range expansion and contraction are crucial in examining the ecological and evolutionary history of both individual species and monophyletic clades (Enserink, 1999; Engler et al., 2004; Gurevitch and Padilla, 2004; Wilson et al., 2004). The importance of preserving geographic ranges of modern species is echoed throughout the modern biological conservation literature (Peterson and Vieglais, 2001; Johnson et al., 2004; Rushton et al., 2004; Thomas et al., 2004; Wilson et al., 2004). The long-term (thousands to hundreds of thousands of years) impact of expansions or contractions in geographic range (those effects requiring hundreds or thousands of years to manifest) can only be directly assessed by studying similar events in the geological past.

5.2   QUANTIFYING GEOGRAPHIC RANGES OF FOSSIL TAXA

Determining the geographic limits of taxon ranges has historically been a key feature of palaeobiogeography. Numerous studies have examined the geographic regions occupied by higher taxa, such as families and orders. Palaeobiogeographic analyses of this type commonly involve plotting taxonomic occurrences on continent or basin scale maps, and examining changes in geographic patterns at the temporal scale of stages (often 5–10 million years in duration) or greater. Biotic comparison based on presence or absence of taxa and perceived biogeographic barriers have contributed enormously to biogeography by documenting first-order biogeographic patterns in the fossil record through the delineation of biogeographic realms and provinces (Boucot et al., 1969; Boucot, 1975; Oliver, 1976; Webby, 1992).

Many palaeobiogeographic analyses concentrate on statistical patterns of similarity between areas using comparative methods, in which the overall similarity of all taxa within regions are considered together rather than investigating the geographic limits of single taxa. Methods for quantifying biogeographic differences range from studies based on similarity metrics and correspondence analysis (Lees et al., 2002; Shen and Shi, 2004) to those in which parsimony is used to determine nested patterns of endemism (Parsimony Analysis of Endemicity; Rosen and Smith, 1988). These types of analyses provide insight into biogeographic patterns operating at ecosystem and biosphere levels. Useful as these methods are, they are designed to identify biogeographic patterns between regions and cannot assess the multidimensional spatial distribution (irregular geographic limits and temporal relationships) of taxa *within* a region.

Species-level analyses provide a framework to assess the relationship amongst biogeographic, palaeoecological, and evolutionary patterns (Adrain et al., 2001). Analyses of species-level patterns would be prohibitively cumbersome if not for increases in both the availability of palaeontological databases (e.g., the Paleobiology Database, http://paleodb.org/cgi-bin/bridge.pl) and computing capabilities in the past decade. Recent studies have incorporated inferred geographic limits of individual species in several contexts. Roy et al. (2001, 2004) determined the latitudinal range of modern and Pleistocene Pacific coast bivalve species to examine range shifts in response to climatic changes. Rode and Lieberman (2004) and Stigall Rode and Lieberman (2005b) reconstructed the ranges of individual brachiopod species across the Late Devonian mass extinction interval and found a significant biogeographic difference between victims and survivors. A similar biogeographic bias in range size and extinction also occurs in Late Triassic ammonoids (W. Kiessling, pers comm.).

5.3 GEOGRAPHIC INFORMATION SYSTEMS IN PALAEOBIOGEOGRAPHY

5.3.1 PALAEOBIOGEOGRAPHIC POTENTIAL OF GIS

Simultaneous analysis of multiple species' ranges can be readily accomplished within a GIS. GIS-based analysis has only recently been applied to palaeobiogeographic analyses, but it offers significant advantages over previous approaches for reconstructing ranges of fossil taxa (Stigall Rode, 2005b). Geographic information systems function as a series of digital maps in which different datasets are stored as independent layers (Theobald, 2003). GIS range maps are dynamic; layers can be shown or hidden and data updated instantly. The ability to efficiently and accurately update data within distribution maps is a significant advantage over traditional range creation methods. Furthermore, GIS databases can be designed to incorporate both temporal and spatial variability (Berry, 1995; Chou, 1997; Burrough and McDonnell, 1998; Stigall Rode, 2005b). This creates a quantitative framework in which to examine temporal and spatial patterns in the fossil record in a statistically rigorous manner. Moreover, creating GIS-based range maps requires a series of species occurrence points, georeferenced to latitude and longitude values. This type of data can potentially integrate seamlessly with large database projects, such as the Paleobiology Database. Species occurrence data can be both extracted from and donated to such databases, which ensures further utility of these data to other researchers.

The combination of these features indicates that GIS analysis of species ranges can result in the creation of hypotheses based on large datasets that can be tested statistically, a fundamental advance for palaeobiogeography. For example, testable hypotheses could be generated examining the relationship of geographic ranges to environmental variables or patterns of speciation and extinction (Stigall Rode, 2005b). In addition, the use of GIS methodology is widespread amongst biologists and ecologists; therefore, incorporation of GIS into the palaeontology 'toolbox' will further enhance opportunities for neo- and palaeobiogeographers to interact. Statistical analyses comparable to those used within the modern biota can then be applied to the fossil record. The use of parallel methods with modern and fossil taxa creates a common framework to compare the dynamics of shifting biogeographic ranges across multiple timescales.

5.3.2 OVERVIEW OF GIS ANALYSES IN PALAEONTOLOGY

Although many biological and geological subdisciplines routinely incorporate GIS into analyses, palaeontological use of GIS methods has been somewhat limited to date. The primary use of GIS in palaeontology has been to create geospatial databases of fossil localities (Benton et al., 2001; Ferguson et al., 2001; Lacruz et al., 2003; McKinney et al., 2003, 2004). These digital databases provide a tremendous opportunity for geospatial analysis that has not yet been exploited analytically. Spatially arrayed morphological features have also been analysed within a GIS including ontogenetic changes in mammalian molar cusps (Jernvall et al., 2000) and the distinctiveness of ammonite suture patterns (Yacobucci and Manship, 2003; Manship, 2004).

Relatively few studies have utilised GIS for analyses testing palaeobiogeographic hypotheses and most of these have analysed species distributions strictly by spatial analysis of point data. Graham et al. (1996a) implemented the first palaeobiogeographic GIS analysis with the FAUNMAP database. The database included point occurrence data for Cenozoic mammals. Spatial occurrences of taxa were utilised to examine range shifts in relation to climate and to analyse community stability through episodes of climate change (Graham et al., 1996a, 1996b; Graham, 2000). Markwick (2002) utilised species occurrence data from the Cenozoic tetrapod fauna of Europe (digitised as point data overlain on polygon-based climate maps) to discern a relationship amongst climate, biogeography, and species diversity. A similar analysis conducted by Vermeersch (2005) analysed the geographic distribution of hominids during the last glacial cycle compared to climate and vegetation distributions. The most sophisticated use of point occurrence data to date was undertaken by Rayfield et al. (2005) who analysed the spatial and temporal occurrence of terrestrial vertebrates to assess the relative validity of vertebrate biochrons in the Middle to Late Triassic.

One of the most powerful aspects of GIS analysis is the ability to reconstruct geographic ranges in multidimensional space. Species occurrence data can be converted from point data into polygon coverages allowing the size and spatial position of geographic ranges to be quantified and compared using spatial statistics. Such studies have been limited to date, but offer incredible potential for palaeobiogeographic analysis (reviewed in 'case studies' below; Stigall and Lieberman, 2006). The use of computer learning-based genetic algorithms to predict species range limits based on environmental parameters has additional potential to quantify and interpret the causes of species range shifts in the fossil record (Stigall Rode and Lieberman, 2005b).

5.3.3 PRACTICAL CONSIDERATIONS FOR IMPLEMENTING GIS IN PALAEOBIOGEOGRAPHIC STUDIES

The use of GIS in palaeontology offers great potential for increased quantification within palaeontological analyses. GIS provides the capacity to analyse large amounts of species occurrence data and produce quantitatively constrained geographic range reconstruction. The basic steps in GIS range reconstruction are database assembly (including taxonomic, geographic, and stratigraphic information for each specimen), mapping of species localities onto modern continental configurations, rotation of species occurrence data onto palaeocontinental reconstructions, and reconstruction of geographic ranges (Rode and Lieberman, 2004; Stigall Rode, 2005b).

The most important and time consuming of these steps is the first: database creation. GIS is a tool, and like all analytical tools, there are important theoretical issues that must be considered. Spatial analysis routines only examine the data provided, and the quality of the output reflects the quality of the source data.

The primary data required to reconstruct a species' range are the taxon identification, geographic location (in decimal degrees) of collection, and the temporal interval of occurrence. Mapping cannot be completed without these three pieces of information, however, a GIS database will have considerably enhanced utility for querying results if additional information can also be included. Recommended additional information includes detailed stratigraphic information (formation, member, zone), additional geographic information (country, state/province, county, city, locality description), systematic classification (higher or lower level classification than unit of interest), and the reference information (specimen ID number from museum/ field collection).

Species occurrence data can be derived from a number of sources including new field collections, museum collections, literature, or archival palaeontological databases. Depending on the level of temporal or spatial accuracy desired to address a specific question, certain data sources may be more appropriate than others. New field collections can have the highest-level precision, because the investigator can validate the taxon ID, geographic coordinates, and stratigraphic horizon personally. Museum collections provide the opportunity to validate taxon ID (which is critical inasmuch as many collections' identifications are several decades old), but locality and stratigraphic horizon must be assumed to be accurate and often have coarse resolution. Data extracted directly from databases often share the coarse stratigraphic and temporal resolution of museum collections and the investigator must also assume taxon identifications are accurate. Data extracted from the literature must be assumed to contain accurate stratigraphic and locality information but may provide opportunity for verification of species ID if specimens are illustrated or systematic descriptions are included.

Species distribution maps, which provide a basis for quantitative spatial analysis, are relatively simple to create within a GIS. Following database construction, species data can be mapped onto modern continental distributions within a GIS (ArcGIS 9.x [ESRI®, 2006] is currently the industry standard) or rotated onto palaeocontinental positions using programs such as PaleoGIS (Ross and Scotese, 2000) or Point Tracker (Scotese, 2004; Figure 5.1). Once the data are rotated onto the palaeocontinental positions, the rotated point data and palaeocontinental maps can be imported into ArcGIS for manipulation. Range maps can then be constructed for each species during each time slice in which it was extant by digitising a polygon to enclose the distribution data for each species as illustrated in Figure 5.1c. More detailed discussion of methods and a stepwise guide to palaeobiogeographic reconstruction methods are published elsewhere (Stigall Rode, 2005b; Stigall, 2006a).

As with the construction of the occurrence database, practical considerations are important in range reconstruction. Multiple techniques could be envisioned for digitising polygons, so investigators should explicitly define the criteria followed when presenting results. It is suggested that when digitising polygons, a specific policy of polygon assembly be adopted by an investigator and that policy should be articulated in resulting publications. Aspects of the policy should include: (1) what distance buffer will be used around occurrence points; and (2) how are the outer boundaries of

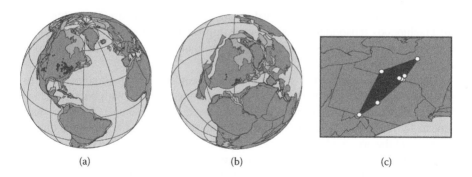

(a)   (b)   (c)

FIGURE 5.1   Steps in GIS range reconstruction. (a) Distribution of all data points plotted onto a modern continental configuration. (b) Reconstruction of all data points present for a single biostratigraphic zone. (c) Reconstruction of the geographic range of the bivalve, *Leptodesma (Leiopteria) nitida* in the *linguiformis* zone (terminal Frasnian conodont zone), range is 30.0×10^3 km². (Modified from Rode, A.L. and Lieberman, B.S. (2005), *Journal of Paleontology*, 79: 267–276.).

the polygon determined (e.g., establishing a polygon with the smallest perimeter, the smallest internal area, or some other criterion). Parsimony principles should be followed where possible to reduce ad hoc assumptions. For example, the reconstructed polygon illustrated in Figure 5.1c in which some occurrences fall within the range is more parsimonious than an alternative polygon in which the range perimeter follows the outline of all occurrences.

5.3.4   Effects of Taphonomic Biases on Palaeobiogeographic Patterns

Of particular concern when reconstructing geographic ranges of fossil species are issues related to taphonomy and the fidelity of the fossil record. Although taphonomic filtering can affect the fossil record in numerous ways (see review in Kidwell and Flessa, 1996), the most significant taphonomic biases to consider within palaeobiogeographic studies are: (1) undersampling (e.g., species was present but not collected at a locality/strata); (2) out-of-habitat transport of specimens; and (3) secular changes in rock volume or facies in outcrop belts. Each of these processes can potentially result in incomplete or inaccurate biogeographic reconstructions if overlooked, but each can also be addressed or mitigated.

The potential for undersampling to affect palaeobiogeographic reconstructions is primarily a function of the abundance of a taxon within the ecosystem. Abundant taxa are more likely to be frequently collected and, consequently, reconstructed ranges for these taxa are more likely to approximate reality. For abundant organisms with strong skeletal elements, such as shallow marine bivalves, analyses have shown that up to 85% of species are preserved in the fossil record (Valentine, 1989). Detailed palaeobiogeographic analyses are, therefore, best suited to common or abundant taxa whose fossil records will exhibit high fidelity. The temporal and spatial limits of an investigation should be scaled appropriately to address this concern. For example, members of Devonian brachiopod species are much more abundant than Devonian phyllocarids (a group of predatory crustaceans); consequently, GIS-based analyses of these groups have been performed on differing spatial and temporal scales: fine scale for brachiopods (conodont zones and township level; Rode and Lieberman, 2004) and a coarser scale for phyllocarids (stages and county level; Rode and Lieberman, 2005), respectively.

Out-of-habitat transport has the potential to confound biogeographic patterns, particularly in studies attempting to link palaeobiogeography with palaeoecology. Studies of modern environments (both marine and continental), however, have repeatedly found that skeletal remains of organisms are rarely transported outside the range of the original habitat of the source population (Kidwell and Bosence, 1991; Kidwell and Flessa, 1996). The primary exceptions to this occur in environments with steep depositional slopes or episodic pulses of high energy which can transport large amounts of skeletal material (Kidwell and Bosence, 1991). Being aware of the depositional slope and energy regime of the sedimentary units under investigation will allow this potential bias to be removed or accounted for within analyses.

The most significant and potentially difficult bias to address in palaeobiogeographic studies is the effect of secular changes in outcrop volume and sedimentary environments within the study area. This type of bias, in fact, is likely to be present within all palaeobiogeographic studies that span a significant range in either time or space. It is critical for investigators to consider whether this type of effect is present within their data and then determine what types of data analysis are best suited to remove any taphonomic biases of this sort. Several statistical approaches can be used depending on the type of secular bias including degrading all data to the level of the lowest interval using subsampling routines, examining changes in area extent using relative or normalised rather than absolute areas, or other types of statistical analyses. In some instances, secular changes may not negatively affect an analysis. For example, Stigall Rode and Lieberman (2005b) recovered a pattern in which several species were increasing their geographic ranges even though the total outcrop area was decreasing. This source of potential bias should be addressed in all palaeobiogeographic studies, however, whether or not a direct link seems initially apparent.

5.4   RELATING GEOGRAPHIC RANGE, ECOLOGY, AND EVOLUTION

Characterising the geographic distribution and limits of species' ranges and analysing biogeographic patterns across the entire biota provides a framework in which to examine palaeoecological and evolutionary patterns. Reconstructed range maps provide the basis for identifying expansions, contractions, or lateral shifts in geographic range of an individual species between temporal intervals (Rode and Lieberman, 2004). Species ranges, such as those described above, can be statistically analysed against environmental parameters within a GIS framework (Rode and Lieberman, 2004) including geostatistical analyses. Species-level analyses can be directly related to immigration or speciation events, emigration or extinction events, and phylogenetic hypotheses of evolutionary relationships (Stigall Rode, 2005b; Stigall Rode and Lieberman, 2005a). Combined analyses of temporal changes in numerous species or clades can shed further light on community level patterns.

More advanced GIS-based methods involving range prediction have also been applied to model species ranges in the fossil record based on environmental preferences. Mensing et al. (2000) predicted the location of Pleistocene packrat (*Neotoma*) middens in Nevada by using a weights-of-evidence model based on geology, elevation, and slope aspect. Stigall Rode and Lieberman (2005b), Maguire and Stigall (2009), Dudei and Stigall (2010), and Walls and Stigall (2011) used a genetic algorithm incorporating multiple environmental variables to predict the geographic extent of species ranges. These studies modified an algorithm, GARP (Genetic Algorithm for Rule-Set Prediction), originally developed to approximate the fundamental niche of a species and then predict the geographic ranges of modern species based on ecological parameters (Stockwell and Peters, 1999; Sanchez-Cordero et al., 2004; Stigall Rode and Lieberman, 2005b). The geological record provides ample environmental data for niche modelling analyses. Environmental variables can be readily discerned from sedimentary structures and lithological parameters in the sedimentary record. This type of analysis, therefore, represents another way in which neo- and palaeobiogeographers can use compatible methods to produce testable hypotheses.

Comparison of species-level biogeographic patterns within a species-level phylogenetic framework provides insight into macroevolutionary processes. For example, observed spatial overlap between two species may be due to several causes including shared environmental tolerances or common ancestry (Wiley and Mayden, 1985; Brooks and McLennan, 1991). Comparing the biogeographic pattern with the evolutionary history of a group provides a method to choose between these alternatives (Stigall Rode, 2005b).

Furthermore, analysis of species-level biogeographic patterns within a phylogenetic context provides the ability to discern patterns of speciation by dispersal or vicariance (Lieberman, 2000, 2003). Several methods of phylogenetic biogeography have been designed to address this issue including parsimony-based analyses, such as Primary and Secondary Brooks Parsimony Analysis (BPA; Brooks et al., 2001). Secondary BPA, in fact, is designed specifically to recover episodes of dispersal and area reticulations in the history of a clade. Other methods developed after BPA include cost-matrix analyses such as Dispersal–Vicariance Analysis (DIVA; Ronquist, 1997) and tree reconciliation analyses such as TreeMap (Page, 1994). A newly developed method, which builds on secondary BPA using nodal analyses to examine specific speciation events across three or more clades, known as Phylogenetic Analysis for Comparing Trees (PACT), provides a direct way to assess vicariance and dispersal events at individual phylogenetic nodes (Wojcicki and Brooks, 2005; Brooks and Folinsbee, 2011). This method integrates seamlessly with examination of reconstructed species-level range reconstructions.

In addition, Lieberman-modified Brooks Parsimony Analysis (LBPA; Lieberman and Eldredge, 1996; Lieberman, 2000) analyses faunal patterns of vicariance and geodispersal exhibited by an entire clade or clades. The LBPA method is the only phylogenetic biogeographic method designed to relate recovered biogeographic patterns to identifiable earth history events, specifically whether cyclical events (e.g., relative sea-level change or climatic oscillations) or singular events (e.g., tectonic collisions or other noncyclical change) contribute more profoundly to the development of observed biogeographic patterns (Lieberman, 2000). By integrating individual species patterns from range map reconstructions with clade patterns from LBPA, fine details of faunal dynamics can be teased apart (Stigall and Lieberman, 2006).

5.5 CASE STUDY: SPECIES-LEVEL ANALYSES OF LATE DEVONIAN BIOGEOGRAPHIC PATTERNS

The Late Devonian was a time of profound evolutionary and environmental change associated with the Frasnian–Famennian Biodiversity Crisis, including reduction in speciation rates, increased extinction rates, rampant species invasions, and ecosystem restructuring (Sepkoski, 1986; McGhee, 1996; Droser et al., 2000). The biodiversity crisis may have lasted as long as three million years with a final pulse of more severe extinction in the last few hundred thousand years of the Frasnian. To unravel the faunal dynamics of this complex crisis, it is critical to understand both the spatial and temporal patterns associated with biodiversity decline.

Prior to the mass extinction interval, a dramatic transition occurred from a highly endemic Middle Devonian fauna to a cosmopolitan biota by the late Frasnian (early Late Devonian; Oliver, 1976, 1990; McGhee, 1996). The expansion of geographic ranges and the transition to a Late Devonian cosmopolitan biota has been documented in many taxa including rugose corals (Oliver, 1976, 1990; Oliver and Pedder, 1994), brachiopods (McGhee, 1981, 1996), foraminifera (Kalvoda and Walliser, 1990), fishes (Young, 1987), conodonts (Klapper and Johnson, 1980; Klapper, 1995), trilobites (Feist, 1991), and land plants (Raymond and Metz, 1995). This changing pattern of geographic range, particularly range expansion events analogous to species invasions, during the Middle to Late Devonian transition, has been implicated in species survival during the biodiversity crisis interval (Rode and Lieberman, 2004).

The Late Devonian strata of Eastern North America, in particular, comprise an excellent record in which to examine changing geographic ranges. The extensive sedimentary record of the Appalachian basin was derived from weathering highlands uplifted during the Acadian Orogeny and preserves a detailed history of the Late Devonian (Frasnian/Famennian) mass extinction, widespread range expansion of species, and a dramatic reduction in speciation rates (Dineley, 1984; McGhee, 1996). The primary tectonic developments were related to the Acadian Orogeny on the eastern margin of the continent (Van der Voo, 1988; Dalziel et al., 1994; Murphy and Keppie, 1998; McKerrow et al., 2000; Figure 5.2). Overthrusting during pulses of the Acadian Orogeny affected the relative relief of the intracratonic basins and arches within Laurentia (Quinlan and Beaumont, 1984; Beaumont et al., 1988). Intracratonic arches were uplifted during episodes of tectonic quiescence, isolating the shallow epicontinental marine basins, and subsided during orogenic pulses (Figure 5.3). The relative relief of intracratonic arches directly influenced the potential for marine species to disperse between the shallow marine basins within Laurentia. Consequently, palaeobiogeographic analyses can address questions related to the relative timing of tectonic uplift and sea level changes because the relative separation of sedimentary basins is reflected directly in biotic exchange versus endemism of their associated faunas.

FIGURE 5.2 Reconstruction of the Late Devonian palaeogeography derived using Ross and Scotese (2000). Biogeographic areas considered in the LBPA analysis are marked. Region abbreviations: A—Appalachian Basin; B—sedimentary basins within Bolivia; C—sedimentary basins of Central North America; E—sedimentary basins of Western Europe; M—Michigan sedimentary basin; W—sedimentary basins of Western North America; WC—Western Canadian sedimentary basins. (Reprinted from Stigall Rode, A.L. and Lieberman, B.S. (2005a). *Palaeogeography, Palaeoclimatology, Palaeoecology*, 222: 272-284. With permission from Elsevier.)

The combination of biotic overturn, geographic expansion, tectonic activity, and substantial fossil and sedimentary record make the Late Devonian of Eastern North America an excellent interval in which to implement GIS methods, assess their accuracy, and employ these methods to decipher the faunal dynamics of a key event in Earth's history. The case study presented below uses range reconstructions constrained by GIS or phylogenetic frameworks to address specific aspects of the Late Devonian Biodiversity Crisis.

5.5.1 Impact of Geographic Range on Species Survival

Using a GIS-based analysis of brachiopod and bivalve species from the Middle to Late Devonian in Laurentia, Rode and Lieberman (2004) investigated the relationship amongst biogeography, relative sea level, and environmental changes. For this analysis a database, including over 8,400 species occurrence points spanning 19 conodont zones from the Givetian to early Fammenian, was assembled from museum and (limited) field collections. The ranges of 341 species of the 28 most common brachiopod and bivalve genera of the Middle to Late Devonian in Laurentia (Rode and Lieberman, 2004) were reconstructed using the polygon boundary method discussed and outlined above (Figures 5.1 and 5.4). Species ranges were spatially quantified within temporal bins approximating conodont zones (approximately 0.5

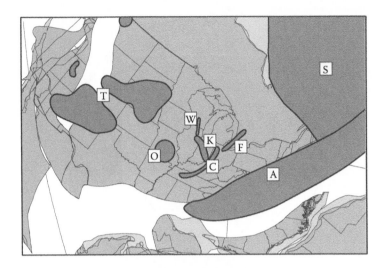

FIGURE 5.3   Intracratonic arches that served as primary barriers to shallow marine dispersal within Laurentia during the Devonian. Abbreviations: A—Acadian highlands; C—Cincinnati Arch; F—Findley Arch; K—Kankakee Arch; O—Ozark Dome; S—Canadian Shield; T—Transcontinental Arch; W—Wisconsin Arch. (Reprinted from Stigall Rode, A.L. and Lieberman, B.S. (2005a). *Palaeogeography, Palaeoclimatology, Palaeoecology*, 222: 272-284. With permission from Elsevier.)

Ma) within a GIS. The timing and extent of species invasions into new tectonic basins were also determined.

Rode and Lieberman (2004) uncovered statistically significant relationships amongst size of species ranges, relative sea-level changes, and species survival through the mass extinction interval. Interbasinal invasion events were not randomly distributed in time but were clustered in three episodes of elevated interbasinal species invasion (Figures 5.4 and 5.5). The timing of these events corresponds to the beginning of the Frasnian, onset of the biodiversity crisis in the mid-Frasnian, and the final stage of the biodiversity crisis in the Late Frasnian. These three pulses of invasion also coincide with transgressive (relative sea-level rise) events IIb, IIc, and IId of Johnson et al. (1985). In addition, species survival through the crisis interval exhibited a clear biogeographic pattern. A clear survival advantage was conferred on species with: (1) larger geographic ranges (*t*-test, $p < 0.001$); (2) one or more episodes of interbasinal invasion in their history (*t*-test, $p \ll 0.001$); or (3) that occupied middle- to outer-shelf environments (X^2 test, $p = 0.004$). Conversely, species that were strictly endemic, had narrow geographic ranges, or lived in nearshore or basinal environments had statistically lower rates of survival into the Famennian.

The relationship between geographic range size and extinction survival was further examined using a genetic algorithm approach to predict the geographic range of a species by modelling the boundaries of its fundamental niche (Stigall Rode and Lieberman, 2005b). This study utilised the GARP modelling system (Stockwell and Peters, 1999), which predicts species ranges by estimating the fundamental ecological requirements (niche) of a species from a set of known occurrence sites each of

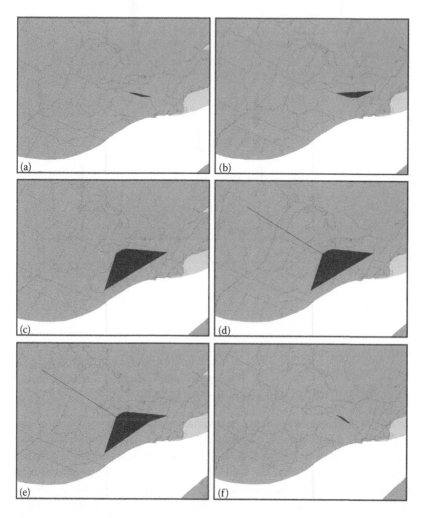

FIGURE 5.4 Palaeobiogeographic ranges of *Schizophoria impressa* in six biostratigraphic intervals of the Middle to Late Devonian indicating episodes of range expansion, contraction, and interbasinal invasion. (a) Lower *falsiovalis* zone, 3.9×10^3 km²; (b) Upper *hassi* zone, 17.8×10^3 km²; (c) *jamieae* zone, 142.4×10^3 km² (range expansion); (d) Lower *rhenana* zone, 143.2×10^3 km² (invasion into Iowa Basin marked by line); (e) *linguiformis* zone, 128.4×10^3 km², (f) Upper *triangularis* zone, 2.6×10^3 km² (post-Late Devonian biodiversity crisis reduction in range). (Modified from Rode, A.L. and Lieberman, B.S. (2004). *Palaeogeography, Palaeoclimatology, Palaeoecology,* 211: 345–359.)

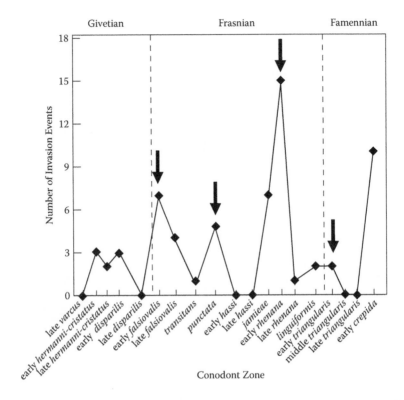

FIGURE 5.5 Intrabasinal invasion events per conodont zone. Vertical arrows indicate the onset of the trangressive events of the Johnson et al. (1985) Devonian sea-level curve. (From Rode, A.L. and Lieberman, B.S. (2004) *Palaeogeography, Palaeoclimatology, Palaeoecology*, 211: 345–359. With permission from Elsevier.)

which is related to a set of environmental variables discerned by sedimentary analysis of the strata from which the fossils were collected. Genetic algorithms provide an alternative to standard regression modelling by including several algorithms in an iterative, artificial-intelligence-based approach. This approach automates decision making by repeatedly analysing a series of local rules that combine categorical, range-type, and logistic rules to obtain higher significance levels than global rules, such as those applied in regression modelling (Stockwell and Peters, 1999; Stockwell and Peterson, 2002). The computer-learning based approach allows the GARP modelling system to customise the set of rules utilised in estimating the ecological niche for each species (Peterson and Vieglas, 2001). The maximum geographic range for the species is then predicted to occupy regions within the study area that satisfy the conditions of the optimised rule set (Stockwell and Peters, 1999; Stigall Rode and Lieberman, 2005b). Niche modelling provides a method to directly utilise sedimentary parameters to predict the geographic region of a species' fundamental niche.

GARP models were based on the dataset of Rode and Lieberman (2004); this allows direct comparison of niche-based range prediction and ranges reconstructed using the GIS method detailed above. GARP and GIS range reconstructions agree

closely (Figure 5.6). In most pairwise comparisons of polygon enclosure and GARP
prediction models, the ranges are roughly consistent between the two outputs (Stigall
Rode and Lieberman, 2005b). Often the predicted ranges quite closely match the polygon ranges, for example, in *Cyrtospirifer chemungensis* (Figure 5.6a,b), which may
suggest both that GARP is accurately predicting known ranges and that the method of
collection of data for the GIS polygon enclosure ranges may be sufficient to reasonably
capture the actual species range. Commonly, although not always, the GARP predicted
range encompasses the entire polygon enclosure range and predicts species to occur
in additional areas adjacent to the polygon range, such as in *Praewaagenoconcha speciosa* (Figure 5.6c,d). Because the GARP algorithm-predicted species ranges are based
on a ruleset that has been trained on the data and contains both internal and external

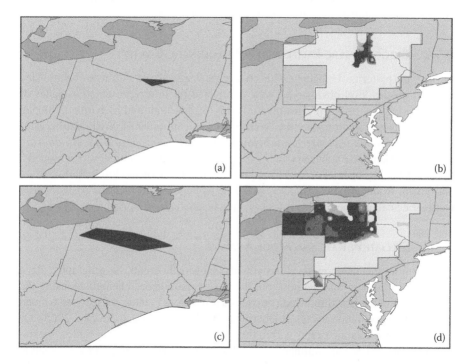

FIGURE 5.6 Comparison of GIS polygon enclosure range reconstructions and GARP distribution predictions. For GARP ranges, black indicates areas where all of the best subset
runs predict a species will occur, and white indicates none of the best runs predicting species
occurrence; the gray scale indicates relative probability between these two end members.
Note that the GIS range reconstructions are rotated onto palaeocontinental positions, whereas
GARP ranges are in modern continental configuration. (a) GIS polygon enclosure range; and
(b) GARP prediction range for the brachiopod *Cyrtospirifer chemungensis* during the *linguiformis* Zone; (c) GIS polygon enclosure range; and (d) GARP prediction range for the
brachiopod *Praewaagenoconcha speciosa* during the *linguiformis* Zone. (From Stigall Rode,
A.L. and Lieberman, B.S. (2005b). Using environmental niche modelling to study the Late
Devonian biodiversity crisis. In *Understanding Late Devonian and Permian-Triassic Biotic
and Climatic Events: Towards an Integrated Approach*. (Eds. D.J. Over, J.R. Morrow, and
P.B. Wignall), Elsevier, Amsterdam, pp. 93–178.)

tests, it should be expected to produce more refined estimates of species range than GIS enclosure models. In fact, predicted ranges that exceed the known species occurrence-bounding polygon provide testable hypotheses for future work in assessing the boundaries of species ranges, predicting ranges of species groups and boundaries of community types, and also determining the quality of the fossil record.

Statistical patterns recovered from GARP analyses are congruent with those from GIS polygon analyses. Large geographic range is statistically associated with species survivorship across the crisis interval for species examined in the *linguiformis* Zone (terminal Frasnian conodont zone; ANOVA, $p = 0.002$; Stigall Rode and Lieberman, 2005b). The relationship of large geographic range and species invasion history with survival through the Late Devonian Biodiversity Crisis recovered in Rode and Lieberman (2004) is again substantiated in this analysis by detailed mapping of species ranges and statistical evaluation of patterns recovered.

5.5.2   BIOGEOGRAPHIC PATTERNS IN A PHYLOGENETIC CONTEXT

To place the patterns of biogeographic change discussed above within an evolutionary context, species-level phylogenies were completed for four representative clades of Middle to Late Devonian marine invertebrates (Stigall Rode and Lieberman, 2005a). The taxa investigated include: the Archaeostraca, a suborder of phyllocarid crustaceans; *Leptodesma (Leiopteria)*, a bivalve subgenus; *Schizophoria (Schizophoria)*, an orthid brachiopod subgenus; and *Floweria*, a genus of orthothetid brachiopods. Species-level phylogenies are published in Rode and Lieberman (2002), Rode (2004), and Stigall Rode (2005a), respectively. These phylogenies formed the basis for speciation rate calculations and species-level phylogenetic biogeographic analysis.

5.5.3   SPECIATION RATE AND MODE ANALYSIS

To determine speciation mode at individual cladogenetic events, biogeographic areas were optimised onto internal nodes in the most parsimonious cladograms using the Fitch Parsimony Algorithm, which provides a framework to interpret speciation mode at individual speciation events (Lieberman, 2002; Wojcicki and Brooks, 2005). Of determinable speciation events within these Middle to Late Devonian clades, speciation by dispersal is the dominant mode (Stigall and Lieberman, 2006). Speciation events due to dispersal comprise 72% of events, whereas vicariance is only implicated in 28% of speciation events (Table 5.1; Rode and Lieberman, 2002; Rode, 2004; Stigall Rode, 2005a). This level of vicariance is greatly reduced compared to that observed in modern clades. Many analyses of speciation mode in the modern biota (Wiley and Mayden, 1985; Brooks and McLennan, 1991, 2002) indicate that the great majority of allopatric speciation occurs via vicariance (72% vicariance, 28% dispersal [cited in Brooks and McLennan, 1991]). Although these studies may underestimate the amount of speciation by dispersal (Cowie and Holland, 2006), other recent studies report similarly high levels of vicariance in marine invertebrate clades (McCartney et al., 2000; Dawson, 2005; Kelly et al., 2006), taxa which would be most comparable to the Devonian study. The reduction in vicariant speciation during this interval may be linked to the marked expansion in geographic range in many species at this time

TABLE 5.1
Episodes of Speciation by Vicariance and Dispersal in Late Devonian Clades

Clade	Number of Vicariance Events	Number of Dispersal Events	Percent Speciation by Vicariance (%)	Percent Speciation by Dispersal (%)
Schizophoria (Schizophoria)[a]	2	17	11	89
Floweria[a]	7	7	50	50
Leptodesma (Leiopteria)[b]	2	6	25	75
Archaeostraca[c]	6	13	32	68
Overall	17	43	28%	72

[a] *Schizophoria (Schizophoria)* and *Floweria* data from Stigall Rode (2005a). *Journal of Systematic Palaeontology*, 3: 133–167.

[b] *Leptodesma (Leiopteria)* data from Rode (2004). *Yale University Postilla*, 229: 1–26.

[c] Archaeostraca data from Rode and Lieberman (2002). *Journal of Paleontology*, 76: 271–286.

(discussed above). Allopatric speciation by vicariance requires the isolation of previously adjacent populations (Mayr, 1942). The numerous range expansion events during this interval, however, would make it difficult to sustain isolation, thereby potentially cutting off the main mechanism of vicariant speciation. In addition, because species with smaller ranges are more likely to become extinct during this interval, the reduction in size from an ancestral range to that of an isolated population is more likely to increase extinction rate, rather than produce a speciation event during this interval.

When the timing of speciation events is constrained, a temporal distinction emerges (Stigall, 2006b, 2010). Species-level phylogenetic hypotheses from Rode (2004) and Stigall Rode (2005a) were converted to strato-cladograms following Smith (1994; Figure 5.7). This method assumes that sister groups diverge at the same time from cladogenetic nodes and creates ghost ranges to connect theoretical speciation and known range data. Rates of biodiversity change, speciation, and extinction were calculated from these data using a pure birth process model. The pure birth process is a deterministic exponential model of taxon growth used for calculating instantaneous rates of biodiversity change within a phylogenetic framework (Hulbert, 1993; Lieberman, 2001; Nee, 2004; Rode and Lieberman, 2005). Rates of biodiversity change were calculated with the following equations, where R is net biodiversity change, S is speciation rate, and E is extinction rate:

$$R = (\ln N_1 - \ln N_0)/\Delta t \qquad (5.1)$$

$$S = (\ln(N_0 + o_0) - \ln N_0)/\Delta t \qquad (5.2)$$

$$E = (\ln(N_0 + o_0) - \ln N_1)/\Delta t \qquad (5.3)$$

In all equations, N_0 is the initial number of species in a clade at time t_0, N_1 is the number of species present at time t_1, Δt represents the duration of the interval $t_1 - t_0$,

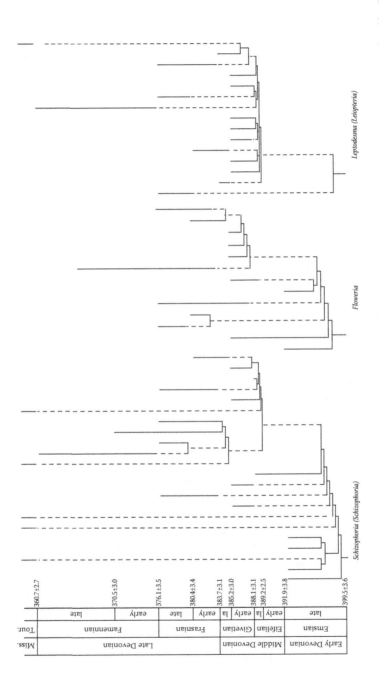

FIGURE 5.7 Species-level phylogenetic hypotheses constrained by stratigraphic occurrence and equal split from phylogenetic node. Absolute ages for Devonian substages from Kaufmann (2006). Phylogenetic data for *Schizophoria* (*Schizophoria*) and *Floweria* from Stigall Rode (2005a) and *Leptodesma* (*Leiopteria*) from Rode (2004); Kaufmann, B. (2006), *Earth Science Reviews*, 76: 175–190; Stigall Rode, A.L. (2005a), *Journal of Systematic Palaeontology*, 3: 133–167; Rode, A.L. (2004), *Yale University Postilla*, 229: 1–26.

and o_0 is the number of speciation events during interval $t_1 - t_0$. Rates of biodiversity change are illustrated in Figure 5.8.

Results of rate calculation indicate several basic patterns. Net biodiversity loss during the Late Devonian biodiversity crisis occurs during the crisis interval (Late Frasnian). This loss is driven partly by extinction (Figure 5.8c), but not entirely. In fact, extinction rates for all clades were higher in preceding intervals than during the Late Frasnian. Speciation rates, however, decline from moderate rates during the Middle Devonian to near zero in the Frasnian (Figure 5.8b). Thus, it is probable that speciation decline was the key determinant in net biodiversity loss. The relative paucity of vicariant speciation discussed above may contribute significantly to the overall pattern of speciation decline because all but two documented vicariance events precede the Late Devonian (Stigall, 2006b).

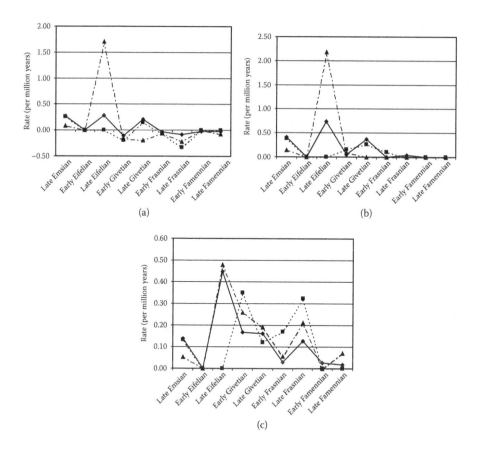

FIGURE 5.8   Instantaneous rates of taxonomic turnover. (a) Rate of net biodiversity change calculated from Equation (5.1), (b) speciation rate calculated from Equation (5.2), (c) extinction rate calculated from Equation (3). Taxon symbols: solid line, *Schizophoria (Schizophoria)*; short dash, *Floweria*; long-short dash, *Leptodesma (Leiopteria)*.

5.5.4 PHYLOGENETIC BIOGEOGRAPHY

To discern the relationship between biogeographic areas occupied by these taxa, LBPA was conducted using the four species-level phylogenies discussed above (Stigall Rode and Lieberman, 2005a). Both the vicariance and geodispersal area cladograms have excellent tree support, but include incongruent biogeographic patterns (Figure 5.9; Stigall Rode and Lieberman, 2005a). Using this method, incongruent patterns are interpreted as relating to causes of singular or noncyclical nature (at least on the time scale of speciation), such as tectonic events (Lieberman and Eldredge, 1996; Lieberman, 2000). Tectonic activity was prevalent in the Middle to Late Devonian of Laurentia and is a likely driver of observed biogeographic patterns. The development, uplift, and downwarp of the intracratonic arches have all been related to orogenic pulses during the Devonian: intracratonic arches experienced uplift during times of orogenic quiescence and downwarp during orogenic pulses (Quinlan and Beaumont, 1984; Beaumont et al., 1988).

The well-resolved vicariance pattern reflects faunal associations known to occur in the Early Devonian faunal provinces (Boucot, 1975; Oliver, 1976) and may reflect Early Devonian separation of basins caused by uplift of tectonic arches during a

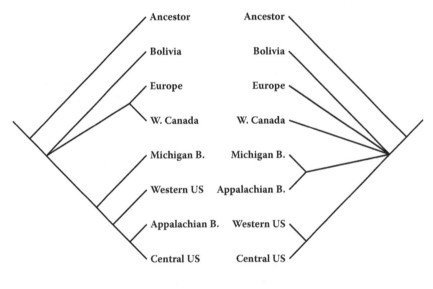

Vicariance Tree   Geo-dispersal Tree

FIGURE 5.9   Strict consensus trees recovered from Lieberman-modified BPA analysis. Vicariance tree indicates the relative order areas were separated. Vicariance tree is the strict consensus of three equally parsimonious resolutions; tree length is 238 steps, with consistency index of 0.85, retention index of 0.51, and g_1 value of –0.54. Geodispersal tree indicates the relative order areas were connected. Geodispersal tree is the strict consensus of eight equally parsimonious resolutions; tree length is 268 steps, with consistency index of 0.84, retention index of 0.42, and g_1 value of –0.55. Both trees rooted with ancestor as the outgroup. (Reprinted from Stigall Rode, A.L. and Lieberman, B.S. (2005a), *Palaeogeography, Palaeoclimatology, Palaeoecology*, 222: 272-284. With permission from Elsevier.)

quiescent phase of the Acadian Orogeny (Ettensohn, 1985; Beaumont et al., 1988). Conversely, the geodispersal cladogram indicates two events which probably occurred following a flexural downwarp of the Findlay and Kankakee arches during the Middle or Late Devonian associated with an Acadian orogenic pulse (Quinlan and Beaumont, 1984; Beaumont et al., 1988; Ver Straeten and Brett, 2000). This offset in perceived timing between vicariance and geodispersal patterns may indicate a fundamental change in the style of biogeographic patterns during the Middle Devonian to Late Devonian. This shift may reflect earlier eustatic versus later tectonic controls on biogeographic patterns, possibly resulting from the relative intensity of Acadian orogenic events.

Comparing the results of the phylogenetic biogeographic analysis with the species-level GIS analysis discussed above, parallel patterns emerge. The Middle to Late Devonian orogenic pulses, which are the likely promoters of the preserved geodispersal patterns, also coincide with transgressive events of the Johnson et al. (1985) sea-level curve. Interbasinal species invasions occur in concert with these transgressive pulses and clearly would have been facilitated by the downwarping of intracratonic arches. This relationship suggests that the interplay of tectonic pulses and transgressive events may have played a major factor in regulating biogeographic patterns and hence biodiversity dynamics during the Late Devonian.

5.6   DISCUSSION

The results of the analyses presented in the case studies above combine to provide a multifaceted insight into the complex interactions operating amongst biogeographic, ecological, and macroevolutionary processes in the time period leading to and including the Frasnian–Famennian biodiversity crisis. Previous studies have noted that although extinction rates were elevated, they were not statistically distinguishable from background rates (Bambach and Knoll, 2001; Bambach et al., 2002); much of the biodiversity lost in this interval must, therefore, be attributable to speciation decline (McGhee, 1996; Bambach et al., 2002). This finding occurs in the case study taxa (Figure 5.8), but the additional species-level analyses reviewed herein allow speciation decline to be examined in detail. Overall speciation decline in these taxa occurs in the Frasnian Stage and is associated with a concomitant decline in the relative amount of speciation by vicariance (Stigall, 2006b).

The case studies presented above illustrate several ways in which quantitative reconstruction of species ranges can be examined in concert with species-level phylogenetic hypotheses to examine the relationship between species ranges and faunal dynamics during the Late Devonian Biodiversity Crisis. In particular, these analyses illustrate that creation of a large set of geographic ranges (in this example with brachiopod and bivalve species) can facilitate examination of changes in geographic range with respect to environmental parameters, such as sea level, as well as the correlation of mass extinction survival with larger geographic ranges and a history of interbasinal invasion events (Rode and Lieberman, 2004; Stigall Rode and Lieberman, 2005b). Results of coordinating analyses of species-level phylogenetic biogeography further corroborate the importance of geodispersal (including interbasinal invasions) and a diminished role for vicariance in driving biogeographic patterns during the Late Devonian (Stigall Rode and Lieberman, 2005a). By combining these results, an understanding of faunal dynamics begins to emerge for the Late Devonian Biodiversity Crisis, one in which species originating from dispersal events and participating in later episodes of interbasinal invasion are successful species that survive the biodiversity crisis interval, whereas stenotopic species with narrow geographic ranges become extinct and do not produce successful daughter species because of the general shutdown of vicariant speciation during this interval. This combination of quantitative and spatial analysis could not have been created without the use of GIS to map species ranges over short temporal intervals.

5.7   CONCLUSIONS

The application of GIS methods in palaeobiogeography offers a powerful technique for the reconstruction of the geographic ranges of species and higher taxa. Data derived from GIS reconstructions can produce datasets amenable to statistical analyses and hypothesis testing. The ability to quantify the spatial and temporal extent and variability in the geographic range of taxa further provides a framework in which geographic changes can be compared with environmental or climatic fluctuations or tectonic events. Hypotheses of relationships between biogeographic patterns and evolutionary and ecological processes can be rigorously assessed.

Results from analyses of Late Devonian species ranges have uncovered relationships amongst species invasions, geographic range, speciation, and extinction during this interval. In particular, species with larger ranges and invasive histories were more likely to survive the biodiversity crisis than species with narrow ranges or lacking prior invasions. Species invasion events were facilitated by tectonic activity and events of relative sea level rise. The survival advantage conferred to invasive (or dispersing) species results in reduced opportunities for speciation via vicariance and also a dramatic decline in overall speciation rate during the Late Devonian. The complex relationship between invasive spread of species and decline in overall speciation rate could not have been discerned without a combination of GIS-based species range reconstructions and well-constrained phylogenetic hypotheses. The complex interplay between biogeographic and evolutionary patterns during the Late Devonian is potentially relevant to the modern biodiversity crisis. If the spread of modern invasive species results in a similar feedback loop between speciation and extinction, we may expect a dramatic decline in speciation in the near future.

The case studies reviewed above represent the first analytical work accomplished through GIS mapping of marine invertebrate species. There is much room to grow with broader use and continued development of more sophisticated GIS methods and applications. The results from these case studies suggest that further GIS-based biogeographic studies coupled with phylogenetic methods have excellent potential to contribute significantly to our understanding of the co-evolution of the Earth and its biota.

ACKNOWLEDGEMENTS

I would like to thank Paul Upchurch, Alistair J. McGowan, and Claire S. C. Slater for organising this volume and the invitation to participate in both the conference and volume. Thanks to Paul Upchurch, Dan Brooks, and one additional reviewer for comments on earlier versions of this chapter. This research was supported by the Ohio University Department of Geological Sciences and Office of International Programs.

LITERATURE CITED

Adrain, J.M., Edgecombe, G.D., and Lieberman, B.S. (2001) Paleobiology; An empirical historical science. In *Speciation in the Fossil Record* (eds. J.M. Adrain, G.D. Edgecombe, and B.S. Lieberman), Kluwer Academic Press, New York, pp. 1–6.

Bambach, R.K. and Knoll, A.H. (2001) Is there a separate class of "mass" extinctions? *Geological Society of America, Abstracts with Programs,* 109: A141.

Bambach, R.K., Knoll, A.H., and Sepkoski, J.J., Jr. (2002) Anatomical and ecological constraints on Phanerozoic animal diversity in the marine realm. *Proceedings of the National Academy of Sciences,* 99: 6854–6859.

Beaumont, C., Hamilton, J., and Quinlan, G. (1988) Orogeny and stratigraphy; Numerical models of the Paleozoic in Eastern Interior of North America. *Tectonics,* 7: 389–416.

Benton, R.C., Evanoff, E., Herbel, C.L., and Terry, D.O.J. (2001) Baseline mapping of fossil bone beds at Badlands National Park. *Technical Report - National Park Service, Geological Resource Division,* 85–94.

Berry, J.K. (1995) *Spatial Reasoning for Effective GIS,* GIS World, Fort Collins, CO.

Boucot, A.J. (1975) *Evolution and Extinction Rate Controls,* Elsevier, Amsterdam.

Boucot, A.J., Johnson, J.G., and Talent, J.A. (1969) Early Devonian brachiopod zoogeography. *Geological Society of America, Special Paper,* 119: 1–113.

Brooks, D.R. and Folinsbee, K.E. (2012) Phylogenetic methods in paleobiogeography: Changing from simplicity to complexity without losing parsimony. In *Palaeogeography and Palaeobiogeography: Biodiversity in Space and Time* (eds. P. Upchurch, A.J. McGowan, and C.S.C. Slater), CRC Press, Boca Raton, FL, pp. 13–38.

Brooks, D.R. and McLennan, D.A. (1991) *Phylogeny, Ecology, and Behavior,* University of Chicago Press, Chicago.

Brooks, D.R. and McLennan, D.A. (2002) *The Nature of Diversity: An Evolutionary Voyage of Discovery,* University of Chicago Press, Chicago.

Brooks, D.R., van Veller, M.G.P., and McLennan, D.A. (2001) How to do BPA, really. *Journal of Biogeography.* 28: 345–358.

Burrough, P.A. and McDonnell, R.A. (1998) *Principles of Geographic Information Systems,* Oxford University Press, Oxford.

Chou, Y.H. (1997) *Exploring Spatial Analysis in Geographic Information Systems,* OnWord Press, Santa Fe, NM.

Cowie, R.H. and Holland, B.S. (2006) Dispersal is fundamental to biogeography and the evolution of biodiversity on oceanic islands. *Journal of Biogeography,* 33: 193–198.

Dalziel, I.W.D., Dalla Salda, L.H., and Gahagan, L.M. (1994) Paleozoic Laurentia-Gondwana interaction and the origin of the Appalachian-Andean mountain system. *Geological Society of America Bulletin,* 106: 243–252.

Dawson, M.N. (2005) Incipient speciation of *Catostylus mosaicus* (Scyphozoa, Rhizostomeae, Catostylidae), comparative phylogeography and biogeography in south-east Australia. *Journal of Biogeography,* 32: 515–533.

Dineley, D.L. (1984) *Aspects of a Stratigraphic System; The Devonian.* John Wiley & Sons, New York.

Droser, M.L., Bottjer, D.J., McGhee, G.R., Jr., and Sheehan, P.M. (2000) Decoupling of taxonomic and ecologic severity of Phanerozoic marine mass extinctions. *Geology,* 28: 675–678.

Dudei, N.L., and Stigall, A.L. (2010) An ecological niche modeling approach to assess bio-geographic response of brachiopod species during the Richmondian (Late Ordovician) Invasion in the Cincinnati Arch. *Palaeogeography, Palaeoclimatology, Palaeoecology,* 296:28–47.

Engler, R., Guisan, A., and Rechsteiner, L. (2004) An improved approach for predicting the distribution of rare and endangered species from occurrence and pseudo-absence data. *Journal of Applied Ecology,* 41: 263–274.

Enserink, M. (1999) Biological invaders sweep in. *Science,* 285: 1834–1836.

Environmental Systems Research Institute (2006) *ArcGIS 9.2.*

Ettensohn, F.R. (1985) The Catskill Delta complex and the Acadian Orogeny; A model. In *The Catskill Delta* (eds. D.L. Woodrow and W.D. Sevon), Geological Society of America, Boulder, CO, pp. 39–49.

Feist, R. (1991) The Late Devonian trilobite crises. *Historical Biology,* 5: 197–214.

Ferguson, C.A., Bodenbender, B.E., Hones, J.L., and Ahmed, K. (2001) Recording the fossil record: A GIS database of Middle Devonian fossils in the Michigan Basin. *Geological Society of America, Abstracts with Programs,* 109: A131.

Graham, R.W. (2000) FAUNMAP Database: Filter effects from field to literature to database to analysis to interpretation. *Geological Society of America, Abstracts with Programs,* 109: A131.

Graham, R.W., Anderson, E., Barnosky, A.D., Burns, J.A., Churcher, C.S., Graham, M.A., Grayson, D.K., Guthrie, R.D., Harington, C.R., Jefferson, G.T., Lundelius, E.L., Jr., Martin, L.D., McDonald, H.G., Morlan, R.E., Schroeder, E.K., Semken, H.A., Jr., Toomey, R.S., III, Webb, S.D., Werdelin, L., Wilson, M.C., and Group, F.W. (1996a) Spatial response of mammals to late Quaternary environmental fluctuations. *Science,* 272: 1601–1606.

Graham, R.W., Graham, M.A., Lundelius, E.L., Jr., and Schroeder, E.K. (1996b) FAUNMAP; An electronic database for mapping changes in the distributions of late Quaternary mammals in the United States. *Abstracts with Programs - Geological Society of America,* 28: 16.

Gurevitch, J. and Padilla, D.K. (2004) Are invasive species a major cause of extinction? *Trends in Ecology & Evolution,* 19: 470–473.

Hulbert, R.C., Jr. (1993) Taxonomic evolution in North American Neogene horses (subfamily Equinae); The rise and fall of an adaptive radiation. *Paleobiology,* 19: 216–234.

Jernvall, J., Kerynen, S.V.E., and Thesleff, I. (2000) Quantifying evolutionary modification of development in mammalian molar topography. *American Zoologist,* 40: 1076–1077.

Johnson, C.J., Seip, D.R., and Boyce, M.S. (2004) A quantitative approach to conservation planning: Using resource selection functions to map the distribution of mountain cari-bou at multiple spatial scales. *Journal of Applied Ecology,* 41: 238–251.

Johnson, J.G., Klapper, G., and Sandberg, C.A. (1985) Devonian eustatic fluctuations in Euramerica. *Geological Society of America Bulletin,* 96: 567–587.

Kalvoda, J. and Walliser, O.H.E. (1990) Late Devonian-Early Carboniferous paleobiogeogra-phy of benthic Foraminifera and climatic oscillations. *Lecture Notes in Earth Sciences,* 30: 183–188.

Kaufmann, B. (2006) Calibrating the Devonian time scale: A synthesis of U–Pb ID–TIMS ages and conodont stratigraphy. *Earth Science Reviews,* 76: 175–190.

Kelly, D.W., MacIsaac, H.J., and Heath, D.D. (2006) Vicariance and dispersal effects on phy-logeographic structure and speciation in a widespread estuarine invertebrate. *Evolution,* 60: 257–267.

Kidwell, S.M. and Bosence, D.W.J. (1991) Taphonomy and time-averaging of marine shelly faunas. In *Taphonomy: Releasing the Data Locked in the Fossil Record* (eds. P.A. Allison and D.E.G. Briggs), Plenum, New York, pp. 115–209.

Kidwell, S.M. and Flessa, K.W. (1996) The quality of the fossil record: Populations, species, and communities. *Annual Review of Earth and Planetary Sciences,* 24: 433–464.

Kiessling, W., and Aberhan, M. (2007) Geographical distribution and extinction risk: Lessons from Triassic/Jurassic marine benthic organisms. *Journal of Biogeography*, 34:1473–1489.

Klapper, G. (1995) Preliminary analysis of Frasnian (Late Devonian) conodont biogeography. *Historical Biology*, 10: 103–117.

Klapper, G. and Johnson, J.G. (1980) Endemism and dispersal of Devonian conodonts. *Journal of Paleontology*, 54: 400–455.

Lacruz, R., Berger, L.R., Brink, J.S., Hancox, P.J., and Ungar, P. (2003) Gladysvale; Fossils, strata, and GIS analysis. *South African Journal of Science*, 99: 283–285.

Lees, D.C., Cocks, L.R.M., and Fortey, R.A. (2002) Quantifying paleogeography using biogeography; A test case for the Ordovician and Silurian of Avalonia based on brachiopods and trilobites. *Paleobiology*, 28: 343–363.

Lieberman, B.S. (2000) *Paleobiogeography: Using fossils to Study Global Change, Plate Tectonics, and Evolution*, Kluwer Academic/Plenum, New York.

Lieberman, B.S. (2001) Analyzing speciation rates in macroevolutionary studies. In *Speciation in the Fossil Record* (eds. J.M. Adrain, G.D. Edgecombe, and B.S. Lieberman), Kluwer Academic Press, New York, pp. 323–339.

Lieberman, B.S. (2002) Phylogenetic biogeography with and without the fossil record; Gauging the effects of extinction and paleontological incompleteness. *Palaeogeography, Palaeoclimatology, Palaeoecology*, 178: 39–52.

Lieberman, B.S. (2003) Paleobiogeography: The relevance of fossils to biogeography. *Annual Review of Ecology, Evolution, and Systematics*, 34: 51–69.

Lieberman, B.S. and Eldredge, N. (1996) Trilobite biogeography in the Middle Devonian; Geological processes and analytical methods. *Paleobiology*, 22: 66–79.

Lomolino, M.V., Riddle, B.R., and Brown, J.H. (2006) *Biogeography*, 3rd edition, Sinauer, Sunderland, MA.

Maguire, K.C., and Stigall, A.L. (2009) Distribution of fossil horses in the Great Plains during the Miocene and Pliocene: An ecological niche modeling approach. *Paleobiology*, 35:597–611.

Manship, L.L. (2004) Pattern matching: Classification of ammonitic sutures using GIS. *Palaeontologica Electronica*, 7: 1–15.

Markwick, P.J. (2002) Integrating the present and past records of climate, biodiversity, and biogeography; Implications for palaeoecology and palaeoclimatology. In *Palaeobiogeography and Biodiversity Change: the Ordovician and Mesozoic-Cenozoic Radiations* (eds. J.A. Crame and A.W. Owen), Geological Society of London, London, pp. 179–199.

Mayr, E. (1942) *Systematics and the Origin of Species*. Columbia University Press, New York.

McCartney, M.A., Keller, G., and Lessios, H.A. (2000) Dispersal barriers in tropical oceans and speciation in Atlantic and eastern Pacific sea urchins of the genus Echinometra. *Molecular Ecology*, 9: 1391–1400.

McGhee, G.R., Jr. (1981) Evolutionary replacement of ecological equivalents in Late Devonian benthic marine communities. *Palaeogeography, Palaeoclimatology, Palaeoecology*, 34: 267–283.

McGhee, G.R., Jr. (1996) *The Late Devonian Mass Extinction; The Frasnian/Famennian Crisis*, Columbia University Press, New York.

McGhee, G.R., Jr. (2001) The Late Devonian mass extinction. In *Palaeobiology II* (eds. D.E.G. Briggs and P.R. Crowther), Blackwell Science, Oxford, pp. 222–225.

McKerrow, W.S., Ahlberg, P.E., Clayton, G., Cleal, C.J., and Eagar, R.M.C. (2000) The late Palaeozoic relations between Gondwana and Laurussia. In *Orogenic Processes; Quantification and Modelling in the Variscan Belt* (eds. V. Haak, C. Mac Niocaill, O. Oncken, and D. Tanner), Geological Society of London, London, pp. 9–20.

McKinney, K.C., Cobban, W.A., and Phan, N.T. (2003) GIS application of the newly digitized USGS-Denver Cretaceous fossil mollusk collection. *Abstracts with Programs - Geological Society of America,* 35: 278.

McKinney, K.C., Cobban, W.A., and Phan, N.T. (2004) GIS application of the newly digitized USGS-Denver Cretaceous fossil mollusk collection. *Open-File Report - U. S. Geological Survey,* 30.

Mensing, S.A., Elston, R.G., Jr., Raines, G.L., Tausch, R.J., and Nowak, C.L. (2000) A GIS model to predict the location of fossil packrat (*Neotoma*) middens in central Nevada. *Western North American Naturalist,* 60: 111–120.

Murphy, J.B. and Keppie, J.D. (1998) Late Devonian palinspastic reconstruction of the Avalon-Meguma terrane boundary; Implications for terrane accretion and basin development in the Appalachian Orogen. *Tectonophysics,* 284: 221–231.

Nee, S. (2004) Extinct meets extant; Simple models in paleontology and molecular phylogenetics. *Paleobiology,* 30: 172–178.

Oliver, W.A., Jr. (1976) Biogeography of Devonian rugose corals. *Journal of Paleontology,* 50: 365–373.

Oliver, W.A., Jr. (1990) Extinctions and migrations of Devonian rugose corals in the eastern Americas realm. *Lethaia,* 23: 167–178.

Oliver, W.A., Jr. and Pedder, A.E.H. (1994) Crises in the Devonian history of the rugose corals. *Paleobiology,* 20: 178–190.

Page, R.D.M. (1994) Maps between trees and cladistic analysis of historical associations among genes, organisms, and areas. *Systematic Biology,* 43: 58–77.

Peterson, A.T. and Vieglais, D.A. (2001) Predicting species invasions using ecological niche modeling: New approaches from bioinformatics attack a pressing problem. *Bioscience,* 51: 363–371.

Quinlan, G.M. and Beaumont, C. (1984) Appalachian thrusting, lithospheric flexure, and the Paleozoic stratigraphy of the Eastern Interior of North America. *Canadian Journal of Earth Sciences,* 21: 973–996.

Rayfield, E.J., Barrett, P.M., McDonnell, R.A., and Willis, K.J. (2005) A geographical information system (GIS) study of Triassic vertebrate biochronology. *Geological Magazine,* 142: 327–354.

Raymond, A. and Metz, C. (1995) Laurussian land-plant diversity during the Silurian and Devonian; Mass extinction, sampling bias, or both? *Paleobiology,* 21: 74–91.

Rode, A.L. (2004) Phylogenetic revision of *Leptodesma (Leiopteria)* (Devonian, Bivalvia). *Yale University Postilla,* 229: 1–26.

Rode, A.L. and Lieberman, B.S. (2000) Using GIS and phylogenetics to study the role of invasive species in the Late Devonian mass extinction. *Geological Society of America, Abstracts with Programs,* 32: 368.

Rode, A.L. and Lieberman, B.S. (2002) Phylogenetic and biogeographic analysis of Devonian phyllocarid crustaceans. *Journal of Paleontology,* 76: 271–286.

Rode, A.L. and Lieberman, B.S. (2004) Using GIS to unlock the interactions between biogeography, environment, and evolution in Middle and Late Devonian brachiopods and bivalves. *Palaeogeography, Palaeoclimatology, Palaeoecology,* 211: 345–359.

Rode, A.L. and Lieberman, B.S. (2005) Integrating evolution and biogeography; A case study involving Devonian crustaceans. *Journal of Paleontology,* 79: 267–276.

Ronquist, F. (1997) Dispersal-vicariance analysis: A new approach to the quantification of historical biogeography. *Systematic Biology,* 46: 195–203.

Rosen, B.R. and Smith, A.B. (1988) Tectonics from fossils? Analysis of reef-coral and sea-urchin distributions from Late Cretaceous to Recent, using a new method. *Geological Society Special Publications,* 37.

Ross, M.I. and Scotese, C.R. (2000) *PaleoGIS/ArcView 3.5.*

Roy, K., Jablonski, D., and Valentine, J.W. (2001) Climate change, species range limits, and body size in marine bivalves. *Ecology Letters,* 4: 366–370.

Roy, K., Jablonski, D., and Valentine, J.W. (2004) Beyond species richness: Biogeographic patterns and biodiversity dynamics using other metrics of diversity. In *Frontiers of Biogeography: New Directions in the Geography of Nature* (eds. M.V. Lomolino and L.R. Heaney), Sinauer, Sunderland, MA, pp. 151–170.

Rushton, S.P., Ormerod, S.J., and Kerby, G. (2004) New paradigms for modelling species distributions? *Journal of Applied Ecology,* 41: 193–200.

Sanchez-Cordero, V., Munguia, M., and Peterson, A.T. (2004) GIS-based predictive biogeography in the context of conservation. In *Frontiers of Biogeography: New Directions in the Geography of Nature* (eds. M.V. Lomolino and L.R. Heaney), Sinauer, Sunderland, MA, pp. 311–323.

Scotese, C.R. (2004) *Point Tracker,* PALEOMAP Project, University of Texas at Arlington.

Sepkoski, J.J. (1986) Patterns of Phanerozoic extinction: A perspective from global data bases. In *Global Events and Event Stratigraphy in the Phanerozoic* (ed. O.H. Walliser), Springer, Berlin, pp. 35–51.

Shen, S. and Shi, G.R. (2004) Capitanian (late Guadalupian, Permian) global brachiopod palaeobiogeography and latitudinal diversity pattern. *Palaeogeography, Palaeoclimatology, Palaeoecology,* 208: 235–262.

Smith, A.B. (1994) *Systematics and the Fossil Record; Documenting Evolutionary Patterns,* Blackwell Scientific, Oxford.

Stigall, A.L. (2006a) *Getting Started with GIS for Paleobiogeographic Reconstruction: Using Excel, PaleoGIS, Point Tracker, ArcView 3.x, and ArcGIS 9.x.* Electronic book, http://oak.cats.ohiou.edu/~stigall/

Stigall, A.L. (2006b) Faunal dynamics of the Late Devonian Biodiversity Crisis: A complex interplay of speciation, extinction, and biogeographic change. *Geological Society of America, Abstracts with Programs,* 38: xx.

Stigall, A.L. (2010). Speciation decline during the Late Devonian Biodiversity Crisis related to species invasions. *PLoS ONE,* 5(12): e15584.

Stigall, A.L. and Lieberman, B.S. (2006) Quantitative paleobiogeography: GIS, phylogenetic biogeographic analysis, and conservation insights. *Journal of Biogeography,* 33: 2051—2060.

Stigall Rode, A.L. (2005a) Systematic revision of the Middle and Late Devonian brachiopods *Schizophoria (Schizophoria)* and *'Schuchertella'* from North America. *Journal of Systematic Palaeontology,* 3: 133–167.

Stigall Rode, A.L. (2005b) The application of Geographic Information Systems to paleobiogeography: Implications for the study of invasions and mass extinctions. *Paleontological Society Memoirs,* 11: 77–88.

Stigall Rode, A.L. and Lieberman, B.S. (2005a) Paleobiogeographic patterns in the Middle and Late Devonian emphasizing Laurentia. *Palaeogeography, Palaeoclimatology, Palaeoecology,* 222: 272–284.

Stigall Rode, A.L. and Lieberman, B.S. (2005b) Using environmental niche modelling to study the Late Devonian biodiversity crisis. In *Understanding Late Devonian and Permian-Triassic Biotic and Climatic Events: Towards an Integrated Approach* (eds. D.J. Over, J.R. Morrow, and P.B. Wignall), Elsevier, Amsterdam, pp. 93–178.

Stockwell, D.R.B. and Peters, D. (1999) The GARP modelling system: Problems and solutions to automated spatial prediction. *International Journal of Geographical Information Science,* 13: 143–158.

Stockwell, D.R.B. and Peterson, A.T. (2002) Effects of sample size on accuracy of species distribution models. *Ecological Modelling,* 148: 1–13.

Theobald, D.M. (2003) *GIS concepts and ArcGIS methods,* 1st edition, Conservation Planning Technologies, Fort Collins, CO.

Thomas, C.D., Cameron, A., Green, R.E., Bakkenes, M., Beaumont, L.J., Collingham, Y.C., Erasmus, B.F.N., Ferreira de Siqueira, M., Grainger, A., Hannah, L., Hughes, L., Huntley, B., Van Jaarsveld, A.S., Midgley, G.F., Miles, L., Ortega-Huerta, M.A., Peterson, A.T., Phillips, O.L., and Williams, S.E. (2004) Extinction risk from climate change. *Nature,* 427: 145–148.

Valentine, J.W. (1989) How good was the fossil record? Clues from the Californian Pleistocene. *Paleobiology,* 15: 83–94.

Van der Voo, R. (1988) Paleozoic paleogeography of North America, Gondwana, and intervening displaced terranes; Comparisons of paleomagnetism with paleoclimatology and biogeographical patterns. *Geological Society of America Bulletin,* 100: 311–324.

Ver Straeten, C.A. and Brett, C.E. (2000) Bulge migration and pinnacle reef development, Devonian Appalachian foreland basin. *Journal of Geology,* 108: 339–352.

Vermeersch, P.M. (2005) European population changes during marine isotope stages 2 and 3. *Quaternary International,* 137: 77–85.

Walls, B.J., and Stigall, A.L. (2011) Palaeobiogeography of Late Ordovician (Maysvillian) brachiopod species of the Cincinnati Ohio region: A niche modeling approach. *Palaeogeography, Palaeoclimatology, Palaeoecology,* 299:15–29.

Webby, B.D. (1992) Global biogeography of Ordovician corals and stromatoporoids, Vol. 6. In *Proceedings of the International Symposium on the Ordovician System* (eds. B.D. Webby and J.R. Laurie), pp. 261–276.

Wiley, E.O. and Mayden, R.L. (1985) Species and speciation in phylogenetic systematics, with examples from North American fish fauna. *Annals of the Missouri Botanical Gardens,* 72.

Wilson, R.J., Thomas, C.D., Fox, R., Roy, D.B., and Kunin, W.E. (2004) Spatial patterns in species distributions reveal biodiversity change. *Nature,* 432: 393–396.

Wojcicki, M. and Brooks, D.R. (2005) PACT: An efficient and powerful algorithm for generating area cladograms. *Journal of Biogeography,* 32: 755–774.

Yacobucci, M.M. and Manship, L. (2003) Putting ammonites on the map; GIS mapping of constraints and asymmetry in ammonite suture shape. *Abstracts with Programs— Geological Society of America,* 35: 317–318.

Young, G.C. (1987) Devonian palaeontological data and the Armorica problem. *Palaeogeography, Palaeoclimatology, Palaeoecology,* 60: 283–304.

6 A Case Study of the Palaeobiogeography of Early Mesozoic Actinopterygians
The Family Ptycholepidae

Raoul J. Mutter

CONTENTS

6.1 INTRODUCTION

The early Mesozoic has recently received much attention with respect to the recovery of marine life from the greatest ever (end-Permian) extinction event and another major extinction event at the end of the Triassic (overviews in Erwin, 2006; Smith, 2007; and references therein). The Permian/Triassic Boundary (PTB) in particular may play a pivotal role in assessing putative cause-and-effect scenarios during the Palaeozoic–Mesozoic transition (Algeo et al., 2007). However, fish species—normally

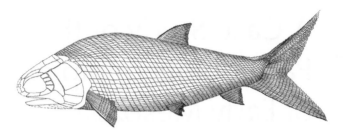

FIGURE 6.1 Sketch of a ptycholepid species from the Hongyanjing Formation in northern Gansu Province (China; possibly Early Triassic in age; see Xu and Gao, 2007).

making up at least 50% of the fossil record of vertebrates and a key to understanding the Palaeozoic–Mesozoic faunal turnover—are remarkably rare in PTB localities, yet quite abundant in a number of classical localities of Late Permian and early Mesozoic age. Due to the scarcity or even absence of fish remains in known PTB sections and rocks immediately on either side of the PTB, these Permian–Triassic localities have recently attracted considerable scientific attention. At the family level (Pitrat, 1973; Schaeffer, 1973; Benton, 1993), the composition of the respective ichthyofaunas seems to be fundamentally different. And indeed, numerous Triassic fish genera cannot be traced back to Palaeozoic ancestors, whereas ichthyofauna diversity increased during the early Middle Triassic (Mutter, 2003).

By reviewing early Mesozoic actinopterygian taxa, and reconstructing their phylogenetic and palaeobiogeographic pathways, much can be deduced that may help to further assess their Palaeozoic ancestry. Using a revised record of a well-known group of actinopterygian fishes, an attempt is made here to reconstruct a possible early Mesozoic evolutionary scenario from a phylogenetic–palaeobiogeographic perspective. Ptycholepidae were defined by Brough (1939). The genera considered to belong to this family appear to be widely distributed in the Induan and persist to the latest part of the Early Jurassic, and probably occur in the Middle Jurassic of central Asia (Su, 1974, 1993) and may persist into the Upper Jurassic of Kazakhstan (Sychevskaya, 2001). Much confusion about ptycholepiform (*sensu* Andrews et al., 1967) interrelationships has arisen from the misinterpretation of the morphological pattern of the scaly lobe found in the caudal fin. In particular, the terms 'abbreviated heterocercal' and 'hemiheterocercal' have been used without formal definition, which led researchers to differentiate *Ptycholepis* Agassiz, 1832 from *Boreosomus* Stensiö, 1921 on erroneous grounds. The phylogenetic interrelationships of Early Mesozoic actinopterygian fishes are still difficult to assess and our understanding is based on very few studies. New species have recently been discovered (Figure 6.1; see also Sychevskaya, 2001; Xu and Gao, 2007) and will provide opportunities for testing phylogenetic/palaeobiogeographic hypotheses and evolutionary scenarios. Currently proposed phylogenies were based on a very incomplete fossil record, and are somewhat subjective (see also Gardiner and Schaeffer, 1989; Gardiner et al., 2005).

A close relationship between acrolepiforms and *Boreosomus* was originally proposed by Gardiner (1967), but has never been tested. Most marine Permian genera described by Aldinger (1937) have never been included in any phylogenetic analysis.

This chapter attempts to close this gap and explore derived ptycholepiform features by comparison with certain Acrolepiformes. The Platysiagiformes, a reasonably well-known and nonspecialised group of Triassic basal actinopterygians, are also included for comparison. The phylogenetic study codes for 23 characters, although 28 characters are identified in Appendix 6A (five characters are currently not applicable). These five characters (12, 17, 20, 23, and 25) will be useful in future studies when applied to a more complete record and a more inclusive sample of genera spanning a greater span of geological time.

Remarkably, many diagnostic characters of individual ptycholepid species turn out to be parsimony uninformative within ptycholepids when put into a larger systematic context. Several causes may account for this problem. First, some characters are poorly known, especially in *Acrorhabdus bertili* and *Boreosomus gillioti* (scored as '?'). Second, the character definitions in this analysis do not capture microscopic skeletal details because ptycholepids are analysed in a larger systematic context and morphological differences between species of ptycholepids may be of minor importance when compared to distantly related taxa. Third, the closest relative(s) of ptycholepids remain unknown, resulting in a disproportionate split between them and all other basal actinopterygians.

Chungkingichthys tachuensis (purportedly quite distinct from all other ptycholepids) was placed in its own family Chungkingichthyidae by Su (1974). In doing so, this species was placed as an even more distant relation due to its stratigraphic position and its 'unknown ancestry' to all other ptycholepids: Su (1974) distinguished Boreolepidae and Ptycholepidae from Chungkingichthyidae, but it can be shown that all five genera share elementary characters as summarised below in the diagnosis of the family Ptycholepidae. Recent review of these genera suggests they may represent a phylogenetic lineage, and that the single family Ptycholepidae (*sensu* Brough, 1939, emended) offers the most coherent grouping for the best-known species (pers. obs.)

The Platysiagidae have been found to share the same stratigraphic range as ptycholepids (Neuman and Mutter, 2005; Mutter, 2005), and are interesting in the context of this study because they can also be traced back to the Early Triassic and serve as an examplar taxon with a similar stratigraphic and palaeobiogeographic distribution.

Various other primitive actinopterygian families 'coexisted' with ptycholepids. Remarkably, many of them can only be traced to the earliest or early Mesozoic and share similar stratigraphic ranges or distributions, but show peculiar specialisations. The Saurichthyidae and the Redfieldiidae share the same stratigraphic range (Induan to Toarcian). The Birgeriidae have a slightly longer range (Induan to Neocomian); the record of Scanilepidae also begins in the Early Triassic, but they are thought to have become extinct in the Rhaetian; the Coccolepidae occur later, in the Sinemurian, and persist at least to the Cenomanian (Schwarz, 1970; Schaeffer et al., 1975; Ørvig, 1978; Rieppel, 1985; Gardiner, 1993; Lund et al., 1995). It seems evident that no close relationships with any of these families exist, although some specific character states were found to correlate with ptycholepids, such as (a) absence of a rostral bone, (b) many postorbitals, and (c) the trapezoidal preoperculum. These features, however, are most probably independently derived—having evolved convergently—because they also occur in many other Mesozoic actinopterygians, and there is very little resemblance in other features and ossification patterns. Hence, none of these families were included in the cladistic analyses but their remotely similar stratigraphic distribution demonstrates the significance of biogeographic considerations with regard to the Early Mesozoic. However, in broader studies, some of the above-mentioned taxa and currently uninformative characters listed in Appendix 6 may prove useful additions for reconstruction of the Early Mesozoic evolution of the more primitive actinopterygians.

6.1.1 PREVIOUS PHYLOGENETIC STUDIES

Early phylogenetic studies considered *Boreosomus* and *Ptycholepis* to be either unrelated or sister groups (i.e., Gardiner, 1960; Véran, 1988). Large-scale studies (i.e., Gardiner and Schaeffer, 1989; Coates, 1999) apparently considered *Boreosomus* to be the best-known ptycholepid genus, despite abundant evidence of *Ptycholepis* from the Middle–Late Triassic. The separation of the two genera, and the choice of *Boreosomus* for all previous phylogenetic analyses, harks back to the erroneously exaggerated difference in morphology between *Boreosomus* and *Ptycholepis*. Historically, *Boreosomus* was interpreted as being closely related to a wide array of more primitive actinopterygians or to the Perleididae on the other hand, whereas Liassic *Ptycholepis* was referred to eugnathids or furids (Brough, 1939; Berg, 1958; Wenz, 1959).

Gardiner (1967) considered his monogeneric family Boreosomidae to be derived from an acrolepid-elonichthyid stock and his Ptycholepiformes (now containing the single family Ptycholepidae, revised) to be an offshoot of the Boreosomidae. However, retaining the family Boreosomidae seems obsolete and the family Chungkingichthyidae finds no convincing support as a well-defined, separate family on the basis of derived characters. In 1984, Gardiner resolved *Boreosomus* as the sister-group of Perleididae and all neopterygians (Gardiner, 1984). Gardiner and Schaeffer (1989) effectively swapped the phylogenetic position of *Boreosomus* with *Australosomus*, and *Boreosomus* (in that analysis being relatively more basal) stood as a sister-group to *Pteronisculus*. More recently, Gardiner et al. (2005) interpreted *Boreosomus* as closely related to *Pteronisculus* and more basal than *Australosomus*, *Perleidus*, and all neopterygians. Coates (1999) clustered *Boreosomus* together with a series of Palaeozoic taxa (including *Coccocephalus*, *Cosmoptychius*, *Mesopoma*, and *Kansasiella*) and *Pteronisculus* and *Birgeria* as a sister group to *Saurichthys*, *Australosomus*, *Perleidus*, and neopterygians. Lund et al. (1995) resolved *Boreosomus* together with *Paleoniscus* and another series of less well-known Paleozoic taxa (including *Cyranorhis*, *Wendyichthys*, and *Rhadinichthys*) as a sister-group to *Pteronisculus*, and the Paleozoic genera *Aesopichthys* and *Canobius*. Other studies have included supposedly close relatives of *Boreosomus* but not *Boreosomus* itself (i.e., Lund, 2000 and Poplin and Lund, 2000). For an overview of these analyses and the most recent analyses of basal actinopterygian interrelationships, see Cloutier and Arratia (2004). Cloutier and Arratia (2004), however, did not include any ptycholepiforms in their study. *Pteronisculus* is the only Triassic genus included in their analyses; it most frequently clusters with the Upper Carboniferous genus *Coccocephalus*.

6.2 MATERIAL AND METHODS

Parsimony analyses were performed using PAUP* 4.0b.10 (Swofford, 2000). Palaeobiogeographic patterns were analysed using a modified Brooks' Parsimony Analysis (BPA) procedure. The approach for optimisation of nodes used for coding in vicariance/geodispersal matrices followed Fitch (1971) and is described in Liebermann and Eldredge (1996) and Lieberman (1997, 2000). The majority of ptycholepid taxa have recently been reviewed, and the following species have been coded for these analyses (ptycholepids in **bold**): *Mimia toombsi* (Gardiner, 1984); ***Acrorhabdus bertili*** (Stensiö, 1921); *Acrolepis sedgwicki* (Agassiz, 1833); *Acropholis stensioei* (Aldinger, 1937); ***Boreosomus arcticus*** (Woodward, 1912); ***Boreosomus gillioti*** (Priem, 1924); *Helmolepis gracilis* (Nybelin, 1977); *Helmolepis cyphognathus* (Neuman and Mutter, 2005); *Hyllingea swanbergi* (Aldinger, 1937); *Platysiagum minus* (Brough, 1939); *Plegmolepis kochi* (Aldinger, 1937); ***Ptycholepis barboi*** (Bassani, 1886); ***Ptycholepis marshi*** (Newberry, 1878); *Pygopterus nielseni* (Aldinger, 1937); *Watsonichthys pectinatus* (Aldinger, 1937); ***Yuchoulepis gansuensis*** (Su, 1993).

6.2.1 PHYLOGENETIC ANALYSES

> Appendix 6B: Data matrix used for the cladistic analysis which is discussed below in detail ('Phylogenetic Analysis'; Figures 6.2, 6.3), containing 16 taxa and 23 (plus 5) characters.

> Appendix 6B: Data matrix used for this cladistic analysis which contained 7 taxa, 13 characters; 6 trees each 20 steps long; consistency index (CI) = 0.85; homoplasy index (HI) = 0.15; CI excluding uninformative characters = 0.81; HI excluding uninformative characters = 0.19; retention index (RI) = 0.73; rescaled consistency index (RCI) = 0.62; unrooted tree(s) rooted using outgroup method; optimality criterion = parsimony. All characters are unordered and equally weighted; character-state optimisation: ACCTRAN.

6.2.2 SYSTEMATIC PALAEONTOLOGY

Class: Osteichthyes (Huxley, 1880)
Subclass: Actinopterygii (Cope, 1887) (*sensu* [Woodward, 1891])
Order: Ptycholepiformes (Andrews et al., 1967)
Family: Ptycholepidae (Brough, 1939)
Genera: *Acrorhabdus* (Stensiö, 1921); *Boreosomus* (Stensiö, 1921); *Chungkingichthys* (Su, 1974); *Ptycholepis* (Agassiz, 1832); *Yuchoulepis* (Su, 1974)

6.2.2.1 Family Diagnosis

Small to large-sized fishes (up to about 450 mm) of fusiform habitus; absence of rostral bone (ch. 1 [1] in cladistic analysis; Figure 6.3); hyomandibula oblique with long and slender processus opercularis; operculum larger than suboperculum; large dermopterotic (= intertemporal) that abuts the nasal bone (ch. 3 [1]); slender-oblong-rectangular frontals (a denotative feature but possibly a primitive retention shared

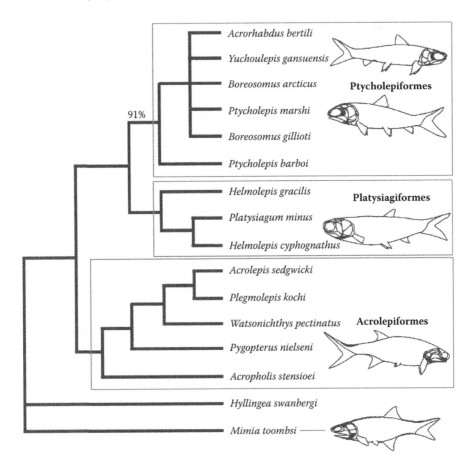

FIGURE 6.2 Consensus tree of 47 trees, 55 steps long. The family Ptycholepidae receives 91% bootstrap support at 2,000 replicates. See the text for discussion.

with other basal actinopterygians; ch. 4 [1], compare Acrolepiformes in Figures 6.2 and 6.3); postorbitals regularly arranged (ch. 14 [1]); dentition weakly developed, teeth small and acutely slender-conical, small molariform within mouth cavity; fins comparatively small and short-based, dorsal fin (rays distally nonbifurcate) adjoins squamation approximately at the level of the ventral fin (in anterior half, ch. 22 [2]), vertical scale rows 20 and 13, respectively; anal fin adjoins at scale row 25–29; caudal fin deeply cleft, ventral lobe in most species slightly better developed (ch. 26 [1]); small fringing fulcra and abbreviate heterocercal caudal fin has enlarged basal fulcra in addition; counts of fin lepidotrichia are: pectoral fin 14–18, ventral fin 11–20; dorsal fin 23–27, anal fin 17–21, caudal fin 50–55; 'palaeoniscoid'-like scales, very low in dorsalmost and ventralmost 10–20 scale rows, square-rectangular in flank region and rhomboidal caudad, usually thick with diversely developed ganoin ornament (ridges, striae, grooves), anastomosing ridges usually end with the posterior ends in the serrations of the posterior margin (see Appendix 6).

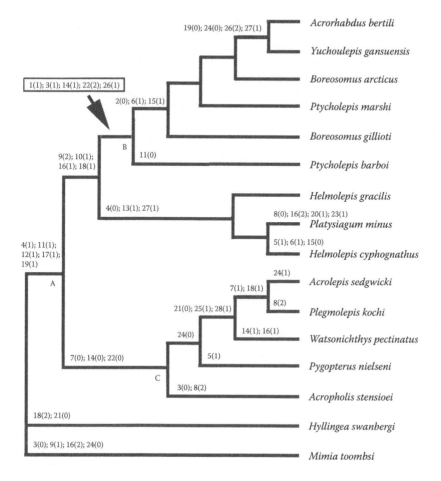

FIGURE 6.3 Phylogenetic hypothesis based on tree number 1 showing the relative position of Ptycholepiformes (Ptycholepidae), Acrolepiformes (node C), and apomorphic characters at respective nodes.

6.2.2.2 Occurrence

Ptycholepids are known from the Lower Triassic Boreal realm (Wordie Creek Formation in East Greenland; Induan [?Griesbachian]; *Boreosomus arcticus* [Woodward, 1912]) to the Lower, possibly Middle Jurassic Shashimiao Formation of Szechuan (Sichuan) basin of central China (*Yuchoulepis gansuensis* [Su, 1993]). However, both earliest and latest occurrences are ambiguous age determinations. None of the specimens of these two taxa has been collected from logged sections and the species from Greenland may be Late Permian, Induan, or Olenekian in age according to the respective authors (Late Permian in Stemmerik et al., 2001; Early Triassic in Mutter, 2003 and Bjerager et al., 2006). The age of the central Chinese species (Szechuan [Sichuan] Basin) may be Early or Middle Jurassic (see Su, 1993; Chang and Miao, 2004). Sychevskaya (2001) reports a Late Jurassic ptycholepid from Eurasia.

6.3   PHYLOGENETIC ANALYSIS

Six Early Triassic to Early Jurassic species of Ptycholepidae, six Late Permian palae-oniscoid actinopterygians, and three Early Triassic to Early Jurassic platysiagids are coded. *Mimia toombsi* (Gardiner, 1984) was chosen as the outgroup. Characters 12 (supplementary rays of gill arches), 17 (presence of presupracleithrum), 20 (presence of 'spinous first' ray in paired fins), 23, and 25 (segmentation of lepidotrichia in dorsal and anal fin) are currently uninformative. All 23 characters used for this analysis are unordered, have equal weight, and are parsimony-informative (matrix in Table 6.1). The DELTRAN option in PAUP (Swofford, 2000) was chosen because it uses a delayed character transformation algorithm and is therefore interpreted to provide more conservative phylogenetic hypotheses, which is consistent with the problematic origin of ptycholepids. Using the TBR branch-swapping algorithm, 47 trees of 55 steps were recovered (the strict consensus tree is shown in Figure 6.2). With regard to the major nodes A, B, and C (Figure 6.3), the majority of trees are similarly resolved but the selected species of Ptycholepidae are variably clustered, not suitable for species diagnoses. Tree number 1 has a CI of 0.67, an HI of 0.33, and an RCI of 0.52. The phylogenetic hypothesis interpreted by tree 1 yields a list of apomorphies, which is suitable for the diagnosis of the family Ptycholepidae (node B), but the diagnoses of species follow the hypothesis based on a reduced dataset (see below). According to tree 1 of this analysis, all ptycholepids (node B in Figure 6.3) lack a rostral (ch. 1 [1]), share a large dermopterotic that abuts the nasal (ch. 3 [1]), have slender, oblong-rectangular frontals (which they share with Acrolepiformes, ch. 4 [1]), their postorbitals form a regular pattern (ch. 14 [1]), their dorsal fin is in the anterior body half (ch. 22 [2]), and their caudal fin in most species is hypocercal (ch. 26 [1]). It is important to note that several other cranial and postcranial features are diagnostic but have no impact on this particular analysis. With respect to ptycholepid intrarelationships and palaeobiogeographical analysis, a more concise set of features has been constructed and is discussed below. A series of alternative topologies and consensus trees is presented in Figures 6.6–6.12.

Although the Ptycholepidae are grouped in this analysis as a family, no consistent interrelationships amongst the genera *Boreosomus*, *Ptycholepis*, and *Chungkingichthys* can be established using this set of characters. Unfortunately, most species of the Ptycholepidae are still too incompletely known to be included in a cladistic analysis. *Ptycholepis barboi* is comparatively well known but all other ptycholepid species included here differ from *P. barboi* in the following respects: they share presence of a postrostral (ch. 2 [0]), the postorbital plate of their maxilla is comparatively small (ch. 6 [1]), and their dermosphenotic is comparatively small (ch. 15 [1]). In *Ptycholepis barboi*, in contrast, the lateral gulars are absent or smaller than the median gular if compared to the condition known in all other ptycholepids (ch. 11[1]).

The Acrolepiformes (node C), including *Plegmolepis kochi*, *Watsonichthys pectinatus*, *Acrolepis sedgwicki*, *Acropholis stensioei*, and *Pygopterus nielseni*, can easily be distinguished from all ptycholepiforms (and platysiagiforms) by the irregular pattern of postorbitals (ch. 14 [0]; except *W. pectinatus* which shows a regular pattern but is still different from all ptycholepids) and the remote position of the dorsal fin (ch. 22 [0]). The opercular region (ch. 7) is somewhat variably developed in these acrolepiforms.

TABLE 6.1
Data Matrix Used in Cladistic Analysis

Species/Character	1	2	3	4	5	6	7	8	9	10	11	12*	13	14	15	16	17*	18	19	20*	21	22	23*	24	25*	26	27	28
Mimia toombsi	0	1	0	0	0	0	2	1	1	0	0	?	0	2	0	2	0	0	0	0	1	1	0	0	0	0	0	0
Acrorhabdus bertili	1	0	1	1	0	1	2	1	2	1	1	1	0	1	1	?	1	1	0	?	?	2	?	?	?	?	?	?
Acrolepis sedgwicki	0	1	2	?	0	0	1	1	2	1	1	1	0	1	0	0	1	1	?	?	?	2	?	1	?	0	0	1
Acropholis stensioei	0	1	0	?	0	0	0	2	0	0	?	?	0	0	0	0	?	0	?	?	1	0	0	1	0	0	?	?
Boreosomus arcticus	1	0	1	1	0	1	2	1	2	1	1	1	0	1	1	1	1	1	1	0	1	2	0	1	1	1	1	0
Boreosomus gillioti	1	0	1	1	0	1	2	1	2	1	?	?	0	1	1	1	1	1	?	?	?	2	?	?	?	?	0	?
Helmolepis gracilis	?	?	?	?	0	0	1/2	1	2	1	1	?	1	1	1	1	1	1	1	?	?	?	0	?	?	0	1	?
Helmolepis cyphognathus	0	1	2	0	1	1	2	1	1	1	?	?	1	2	?	?	?	1	?	0	1	1	0	1	1	0	0	0
Hyllingea swanbergi	?	?	2	0	?	?	?	?	1	1	1	?	?	?	?	0	?	2	0	?	?	?	0	0	0	?	?	?
Platysiagum minus	0	1	2	0	0	0	2	0	0	1	0	0	1	2	0	2	1	1	1	1	0	1	1	1	0	0	0	0
Plegmolepis kochi	0	1	?	1	?	0	1	2	?	0	?	?	0	0	0	0	1	1	1	?	0	?	?	0	1	?	?	?
Ptycholepis marshi	1	0	1	1	0	-	2	1	2	1	1	1	0	1	1	1	1	1	1	0	1	2	0	1	0	1	0	0
Ptycholepis barboi	1	1	1	1	0	0	2	1	2	1	0	?	0	1	0	1	1	1	1	0	1	2	?	1	?	1	0	0
Pygopterus nielseni	?	?	?	?	1	0	0	1	0	0	?	?	0	0	0	1	0	0	0	0	0	1	0	0	0	0	0	0
Watsonichthys pectinatus	?	?	2	1	0	0	0	1	1	0	?	?	0	1	0	0	?	1	1	?	0	?	?	0	?	0	0	1
Yuchoulepis gansuensis	1	0	?	?	0	1	2	1	2	1	?	?	0	1	1	1	1	1	?	0	?	2	0	0	?	2	1	0

[a] Currently (and for this specific analysis) uninformative characters are marked with asterisks.

In the major phylogenetic hypothesis discussed in this chapter (based on tree 1 of 47 trees; Figure 6.3), the order Acrolepiformes (node C) forms a sister-group together with Ptycholepiformes (Ptycholepidae) plus Platysiagiformes (node A), and the very incompletely known, edentulous (ch. 18 [2]) species *Hyllingea swanbergi* (*incertae sedis*) with its long-based pelvic fin (ch. 21 [0]) is resolved as the sister-group of all three orders. According to Schultze and Cumbaa (2001), however, a long-based pelvic fin is a suspected basal osteichthyan feature and may not be 'phylogenetically informative'. Another interesting, probably synapomorphic, ptycholepiform character can be identified when comparing these three orders: there are slender, oblong-rectangular frontals (ch. 4 [1]) in ptycholepiforms, which have apparently been lost in Platysiagiformes.

6.4   DISCUSSION OF PTYCHOLEPID INTRARELATIONSHIPS AND PALAEOBIOGEOGRAPHIC ANALYSIS

One may assume no specific palaeobiogeographic trends can be deduced from the phylogenetic hypothesis in Figure 6.3 using Liebermann's modified BPA approach because the resulting vicariance and geodispersal trees derived from the respective matrices in Table 6.2 are identical. Yet, Lieberman (2000) states that, in the case of identical vicariance and geodispersal trees, cyclical barrier formation and destruction have affected the taxa. However, the three orders of basal actinopterygians are phylogenetically not necessarily directly related, and therefore no palaeobiogeographical signal may be expected. Palaeobiogeography of ptycholepids and their intrarelationships were therefore analysed in a separate study using the six best-known taxa and best-known features only (13 characters redefined in part [Appendix 6B]; data matrix in Table 6.3), *Acropholis stensioei* as outgroup, and the same ptycholepid species. An exhaustive search was performed on this matrix, yielding six trees of 20 steps (CI = 0.81, HI = 0.19; RCI = 0.62). The list of apomorphies of tree 1 provides many diagnostic characters for these species and the well-resolved tree number 4 (Appendix 6B) of this analysis can be used for translation and coding of geodispersal and vicariance matrices (Figure 6.4a, Table 6.4). Areas selected for biogeographic analysis were predetermined by selection of the six most completely known ptycholepid species (see Figure 6.4 and Table 6.5). Both outgroups come from the northern Proto-Atlantic, a potential ancestral region, and are the most closely related or best-known outgroup used in this analysis. The defined palaeogeographic areas (Figure 6.4b) are not necessarily physically delimited but are rather understood as well-separated (distantly placed) palaeoenvironmental realms.

Each species is replaced by an area (rather than a locality), and the nodes are modified using Fitch's (1971) algorithm and Liebermann's (2000) method of vicariance/geodispersal matrix construction (see also Liebermann and Eldredge, 1996). Analysis of the recoded nodes yields differing results, which can be interpreted as evidence of vicariance and geodispersal events, respectively (Figure 6.4b). Using the vicariance matrix, the result is a single most parsimonious and reasonably well-resolved tree (27 steps, CI = 0.74; RI = 0.63; RCI = 0.47). Using the geodispersal matrix yields two most parsimonious and relatively poorly resolved trees (hence, a consensus tree is used for mapping). When plotted as mirror-images (Figure 6.4b), the two resulting

TABLE 6.2A
Geodispersal Matrix[a]

Region/Node	1	2	3	4	5	6	7	8	9	10	11	12	13	14	15	16	17	18	19	20	21	22*	23*	24*	25	26	27*
Outgroup region	1	0	0	0	0	0	0	0	0	0	0	0	0	0	0	0	0	0	0	0	0	0	0	0	0	0	0
Arctic Proto-Atlantic 1 (Palaeozoic)	1	1	0	0	0	0	0	1	1	1	1	1	1	1	1	1	1	1	0	0	0	0	0	0	0	0	0
Western Atlantic basins	0	0	0	0	0	0	0	0	0	0	0	0	0	0	0	0	0	0	0	0	0	0	2	0	0	0	0
Southern Oman-Mozambique Street	0	0	0	0	0	0	0	0	0	0	0	0	0	0	0	0	0	0	0	0	0	0	0	2	0	0	0
Arctic Proto-Atlantic 2 (Mesozoic)	1	1	1	1	0	0	1	1	0	1	0	0	0	0	0	0	0	0	1	0	1	1	1	1	1	1	1
Eastern Paleotethys	0	0	0	0	0	0	0	0	0	0	0	0	0	0	0	0	0	0	0	0	0	0	0	0	0	0	2
Eastern Panthalassa	1	1	1	1	0	1	0	0	0	0	0	0	0	0	0	0	0	0	0	0	0	2	0	0	0	0	0
Northwestern Tethys	1	1	1	1	0	0	1	1	0	0	0	0	0	0	0	0	0	0	1	1	0	0	0	0	0	0	0
Ancestral region	0	0	0	0	0	0	0	0	0	0	0	0	0	0	0	0	0	0	0	0	0	0	0	0	0	0	0

[a] Coded from phylogenetic analysis using a modified version of BPA. Positions 2–8, 10, 19–21, 24, 25, and 27 are treated as 'ordered' positions (*).

TABLE 6.2B
Vicariance Matrix[a]

Region/Node	1	2*	3*	4*	5*	6*	7*	8*	9	10*	11	12	13	14	15	16	17	18	19*	20*	21*	22	23*	24*	25	26	27*
Outgroup region	1	1	0	0	0	0	0	0	0	0	0	0	0	0	0	0	0	0	0	0	0	0	0	0	0	0	0
Arctic Proto-Atlantic 1 (Palaeozoic)	1	2	1	0	0	0	2	1	1	2	1	1	1	1	1	1	1	1	1	1	0	0	0	0	0	0	0
Western Atlantic basins	0	0	0	0	0	0	0	0	0	0	0	0	0	0	0	0	0	0	0	0	0	0	2	2	0	0	0
Southern Oman-Mozambique Street	0	0	0	0	0	0	0	0	0	0	0	0	0	0	0	0	0	0	0	0	2	0	0	2	0	0	0
Arctic Proto-Atlantic 2 (Mesozoic)	1	2	2	2	1	1	1	1	1	1	0	0	0	0	0	0	0	0	2	1	2	1	1	1	1	1	1
Eastern Paleotethys	0	0	0	0	0	0	0	0	0	0	0	0	0	0	0	0	0	0	0	0	0	0	0	0	0	0	2
Eastern Panthalassa	1	2	2	1	1	2	0	0	0	0	0	0	0	0	0	0	0	0	0	0	0	1	0	0	0	0	0
Northwestern Tethys	1	2	2	1	2	1	2	2	1	1	0	0	0	0	0	0	0	0	2	2	1	0	0	0	0	0	0
Ancestral region	0	0	0	0	0	0	0	0	0	0	0	0	0	0	0	0	0	0	0	0	0	0	0	0	0	0	0

a Vicariance matrix after recoding from Figure 6.1. Positions 22–24 and 27 are treated as 'ordered' positions (*).

TABLE 6.3
Reduced Data Matrix[a]

Species/Character	1	2	3	4	5	6	7	8	9	10	11	12	13
Acropholis stensioei	1	0	0	?	?	0	0	2	0	0	1	0	0
Acrorhabdus bertili	0	1	1	0	0	0	0	?	1	0	1	?	0
Boreosomus arcticus	0	1	1	1	1	0	0	0	1	1	0	0	0
Boreosomus gillioti	0	1	1	?	1	0	0	0	1	?	0	0	0
Ptycholepis barboi	1	1	0	1	1	1	1	1	1	2	0	0	1
Ptycholepis marshi	0	2	1	1	1	1	1	1	1	2	2	0	1
Yuchoulepis gansuensis	0	2	1	?	?	1	0	1	1	?	0	1	1

[a] Used for analysis of ptycholepid intrarelationships and further palaeobiogeographic analysis. See the text for discussion.

area cladograms display areas of agreement (or overlap: black clades) but also areas of conflict (grey clades). The resulting discrepancies are restricted to the Triassic, concentrated on three intervals, the Early Triassic, Early–Middle Triassic transition, and Middle–Late Triassic transition, that are probably of palaeobiogeographic relevance.

This pattern of putative vicariance/geodispersal events is also interesting in the light of two well-documented extinction events. Apparently, following the end-Permian extinction event, the neopterygians (modern actinopterygians) diversified. However, the more primitive ptycholepids had a wider geographical distribution (geodispersal) and then radiated during the Middle Triassic (even if the number of genera remained constant or even declined). Previous assessments of diversity of

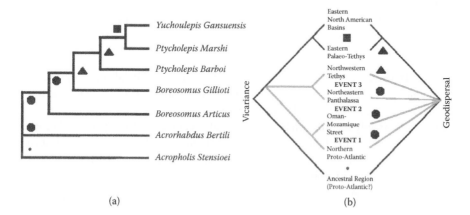

(a)　　　　　　　　　　　　　　　　(b)

FIGURE 6.4 Translation of a selected phylogenetic hypothesis into area cladograms by means of recoded geodispersal and vicariance matrices. (a) Tree number 4 from Appendix 6B, using the reduced set of characters on the six best known ptycholepid species. (b) Vicariance and geodispersal trees combined to show nodes and clades corresponding (black) versus noncorresponding (grey, palaeobiogeographic events 1, 2, and 3 in text), suggesting three possible palaeobiogeographic events. See the text for discussion.

TABLE 6.4A
Geodispersal Matrix[a]

Region/Node	1	2	3	4	5	6	7	8	9	10	11
East Greenland	1	1	1	0	0	0	0	0	0	0	0
Northwestern China	1	0	1	0	1	0	0	0	0	0	0
Northern Madagascar	1	0	1	0	1	1	1	0	0	0	0
Eastern North America	1	0	1	0	1	0	1	1	1	0	0
Western North America	1	0	1	0	1	0	1	0	1	1	0
Central Europe	1	0	1	0	1	0	1	0	1	0	1
Ancestral region	0	0	0	0	0	0	0	0	0	0	0

[a] Recoded for palaeobiogeographic analysis. All positions treated as 'unordered.'

TABLE 6.4B
Vicariance Matrix[a]

Region/Node	1	2*	3	4*	5*	6*	7*	8*	9*	10*	11*
East Greenland	1	2	1	1	0	0	0	0	0	0	0
Northwestern China	1	1	1	2	2	1	1	0	0	0	0
Northern Madagascar	1	1	1	1	2	2	2	1	1	0	0
Eastern North America	1	1	1	1	2	1	2	2	2	1	1
Western North America	1	1	1	1	2	1	2	1	2	2	1
Central Europe	1	1	1	1	2	1	2	1	2	1	2
Ancestral region	0	0	0	0	0	0	0	0	0	0	0

[a] Recoded for palaeobiogeographic analysis. All positions except 1 and 3 are treated as 'ordered' positions (*).

actinopterygians across the PTB (Pitrat, 1973; Schaeffer, 1973) did not recognise this event amongst basal actinopterygians. The Late Permian record of ptycholepid relatives, however, remains virtually unknown. Furthermore, the end-Triassic extinction event is not directly recorded in the ptycholepid fossil record itself, possibly because several members of the family invaded the nonmarine realm before the end of the Triassic, and nonmarine species richness remained high during the Early Jurassic (see Table 6.5). The palaeobiogeographic significance of this group of basal actinopterygians is therefore underlined by their fossil record, which most probably predates and certainly postdates two major extinction events.

Ptycholepids also possess a unique combination of characters, namely a very slender-fusiform body shape with their dorsal fin relatively far anterior and a slender, but powerful, caudal peduncle reminiscent of open marine, pelagic fishes. In combination with their long and slender jaws with feeble dentition it would appear these fishes were predominantly planktoniverous or were preying on small invertebrates lacking thick shells. The latter condition is shared with the platysiagiforms,

TABLE 6.5

Summary List of Currently Described 25 Ptycholepid Species[a]

List of Described Ptycholepid Species	Number of Known Specimens	Geological Age
Boreosomus arcticus (**Woodward, 1912**)	>8	**Early Triassic**
Boreosomus reuterskioeldi (Stensiö, 1921)	>9	Early Triassic
Boreosomus scaber (Stensiö, 1921)	? (scales in shreds)	Early Triassic
Boreosomus gillioti (**Priem, 1924**)	3	**Early Triassic**
Boreosomus sp. nov. (Mutter, subm.)	2	Early Triassic
Acrorhabdus bertili (**Stensiö, 1921**)	5	**Early Triassic**
Acrorhabdus asplundi (Stensiö, 1921)	1	Early Triassic
Acrorhabdus latistriatus (Stensiö, 1921)	1	Early Triassic
Acrorhabdus piveteaui (Nielsen, 1942)	137	Early Triassic
Ptycholepis priscus (Bürgin, 1992)	>10	Middle Triassic
Ptycholepis schaefferi (Bürgin, 1992)	>10	Middle Triassic
Ptycholepis magnus (Bürgin, 1992)	8	Middle Triassic
Ptycholepis viai (Beltan, 1984)	3	Middle Triassic
Ptycholepis barboi (**Bassani, 1886**)	**61**	**Middle Triassic**
Ptycholepis marshi (**Newberry, 1878**)	**>30**	**Late Triassic**
Ptycholepis avus (Kner, 1866)	? (several)	Late Triassic
Ptycholepis raiblensis (Bronn, 1858)	? 1 (possibly several)	Late Triassic
Ptycholepis gracilis (Davis, 1884)	>3	Late T./E. Jurassic
Ptycholepis bollensis (Agassiz, 1832)	14	Early Jurassic
Ptycholepis minor (Egerton, 1852)	>3	Early Jurassic
Ptycholepis monilifer (Woodward, 1895)	1	Early Jurassic
Ptycholepis curta (Egerton, 1854)	>6	Early Jurassic
Chungkingichthys tachuensis (Su, 1974)	5	Early Jurassic
Yuchoulepis szechuanensis (Su, 1974)	6	Early Jurassic
Yuchoulepis gansuensis (**Su, 1993**)	**16**	**Early Jurassic**

[a] Coded species in bold.

and possibly shared with certain acrolepiforms (see Aldinger, 1937). Recent research on chondrichthyans has yielded evidence that the end-Permian extinction event may have been rather 'selective', possibly enhancing survival or extinction, in relation to certain biological specialisations leading to ecological adaptations (Mutter and Neuman, 2006, 2008a, 2008b, 2009). At present, it is unclear what specific cause and effect scenarios took place at the two 'extinction boundaries', and it is evident that only more detailed fieldwork at suitable boundary sections and investigations into latest Palaeozoic and early Mesozoic fish groups will enable us to track putative phylogenetic lineages and palaeobiogeographic patterns across both boundaries in order to better tackle extinction/recovery scenarios.

At present, it is noteworthy that a poor marine ptycholepid record and identical vicariance/geodispersal trees mark both the end-Permian and end-Triassic extinction events. Palaeobiogeographical events during these times may indeed be linked to cyclical barrier formation and destruction (according to Liebermann, 2000), but testing this interpretation with respect to current extinction scenarios taking place at these boundaries is beyond the scope of this chapter. Also, the palaeobiogeographic patterns of many other reasonably well-known actinopterygians spanning this timeframe remain to be studied.

6.5   CONCLUSIONS

Early Mesozoic actinopterygian phylogeny and palaeobiogeography may be successfully studied using a combined cladistic/modified BPA approach applied to the best-known taxa only. The vicariance tree differs from the geodispersal tree in being better resolved, and the differences in tree topology relate to three nodes, which are interpreted (as stated above) to be of palaeobiogeographic significance (Figures 6.4, 6.5): (event 1), a biogeographical event during the Early Triassic (northern Proto-Atlantic–Oman-Mozambique); (event 2), a biogeographical event during the Early-Middle Triassic (Oman-Mozambique–northeastern Panthalassa); (event 3), a possible Middle-Late Triassic biogeographical event (northeastern Panthalassa–northwestern Tethys). Yet no differing vicariance/geodispersal trees have been obtained for ptycholepids during the Late Triassic–Early Jurassic transition (eastern Palaeotethys–eastern North American basins) or near their assumed origin in the Late Permian Proto-Atlantic (*). Episodes of cyclical barrier formation and destruction around the time of Permian/Triassic and Triassic/Jurassic boundaries are thus suspected. Figure 6.5 shows a simplified mapping of the currently known record of ptycholepids on a palaeogeographic map, and these interpretations of the vicariance/geodispersal trees seem plausible, with east–west and north–south geodispersal across all Pangaea (events 1 and 2 above) which must have taken place in the Early Triassic (Induan, polygonal symbols, and bright grey arrows), whereas predominantly east–west distribution across central Pangaea (probable vicariance, event 3 above) occurred after the Middle Triassic (triangle and square symbols, dark grey and black arrows).

Direct evidence for geodispersal during the Early Triassic is actually found in the fossil record with specimens of *Boreosomus* cf. *B. arcticus* from the Induan to Smithian of the northern Proto-Atlantic (localities in Spitzbergen and northwestern Canada). The fact that no geodispersal (or vicariance) events can be identified during the Late Triassic–Early Jurassic transition is either an effect of less complete preservation of the rock or fossil record during that time, or is an indication of increased structural complexity of the marine shelf. Although latitudinally less abundant, ptycholepids invaded mixed waters and freshwater-influenced habitats toward the end of the Triassic, whereas the family seems restricted to the marine realm during the Early–Middle Triassic. Periodic environmental change may have opened up mixed water bodies at mid-latitudes during that time. Yet such an interpretation is unlikely to find support using a method distinguishing long-term vicariance and geodispersal events with a relatively small but well-resolved data matrix, and may have to be inferred using the more complete fossil record of several fish groups and further consideration of the palaeoecological role of these fish groups.

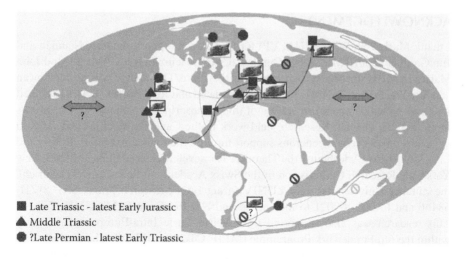

■ Late Triassic - latest Early Jurassic
▲ Middle Triassic
● ?Late Permian - latest Early Triassic

FIGURE 6.5 Reconstruction of Pangaea at about 230 Ma before present (Late Triassic; modified from Golonka and Ford, 2000). This figure is greatly simplified to enhance clarity, but is based on the complete fossil record known ('total evidence') and illustrates possible paths of biogeographical events (geodispersal) in the Early Triassic (polygonals and bright grey lines), a biogeographic event (geodispersal) during the Early–Middle Triassic (triangles and black lines) and possible vicariance events in the Middle–Late Triassic (squares and dark grey lines) (scale: fragmentary record; full circles: complete specimens; strike-through circles: probable absence; double-headed arrows: unknown geodispersal; asterisk: earliest record in ?Late Permian). Symbols: Polygons (Early Triassic): based on Woodward (1912), Stensiö (1918, 1921, 1932), Priem (1924), Chabakov (1927), Corroy (1928), Brough (1933), Nielsen (1942), Gardiner (1966, 1967), Beltan (1968, 1972, 1984, 1996), Lehman (1952, 1966), Schaeffer and Mangus (1976), and Ørvig (1978). Triangles (Middle Triassic): based on de Zigno (1891), Bassani (1886), Abel (1906), Alessandri (1910), Andersson (1916), Deecke (1926), Bürgin (1992), and Mutter (2002). Squares (Late Triassic—Early Jurassic): based on Agassiz (1834, 1844), Bronn (1837, 1858), Williamson (1849), Egerton (1852, 1854), Morris (1854), Quenstedt (1858, 1867), Cotteau (1865), Kner (1866, 1967), Sauvage (1875, 1891), Newberry (1878, 1888), Davis (1884), Zittel (1890), Woodward (1895, 1896), Scupin (1896), Eastman (1905), Hay (1902, 1930), Hussakof (1908), Hennig (1918), Gross (1935), Wenz (1959, 1967), Hauff (1960), Gardiner (1960, 1967), Su (1974, 1993), Schaeffer *et al.* (1975), and Lamaud (1977). (Modified from Golonka, J. and Ford, D. (2000), *Palaeogeography, Palaeoclimatology, Palaeoecology,* 161: 1–34.

Ptycholepids are a family of relatively easily recognisable actinopterygians with peculiar specialisations and limited stratogeographic provenance, and their occurrence is likely to be tightly linked to specific environmental events during the early Mesozoic. It will be interesting to discover to what degree the palaeobiogeographic patterns of this actinopterygian lineage are matched or contradicted by the fossil record of other early Mesozoic actinopterygians. Reconstructing phylogenetic and palaeobiogeographic patterns in Early Mesozoic fishes will shed light on possible relationships across the Permian/Triassic and Triassic/Jurassic boundaries, whose importance in assessing and understanding the Palaeozoic–Mesozoic faunal turnover has been widely recognised.

ACKNOWLEDGEMENTS

I thank Mark V.H. Wilson (UALVP) for continuous support, Andrew Neuman and Jim Gardner (both TMP), Steve Cumbaa (CMN), Heinz Furrer (PIMUZ), and John Maisey (AMNH) for access to specimens in their care. The input of Alistair McGowan and Stig Walsh (both NHM) is greatly appreciated and John Bruner (UALVP) helped with comments on an earlier version of this manuscript. Two anonymous reviewers improved this chapter. Research, field work in the Canadian Rockies and visits to museums have received generous support through the foundation of Dr. Robert and Lina Thyll-Dürr (Arlesheim), the Theodore Roosevelt Memorial Fund (AMNH, New York), the Research Commission of the Swiss Academy of Sciences (ETH Zürich), the Zürcher Universitätsverein ZUNIV (grant FAN), research grants SNF 81ZH-68466 and PA002-109021 to the author, and NSERC Grant to Mark V.H. Wilson. This research was also supported by a Marie Curie Intra-European Fellowship within the 6th Framework Programme (MEIF-CT-2006-023691). The final stage of this publication was supported through a Postdoctoral Research Fellowship at the BPI for Palaeontological Research, University of the Witwatersrand, Johannesburg, South Africa.

APPENDIX 6A: LIST OF CHARACTERS USED IN PHYLOGENETIC ANALYSIS

Characters modified from Wenz (1959, 1967), Schaeffer et al. (1975), Lund et al. (1995), and Mutter (2002).

Character 1: Rostral
    0, present; 1, absent or fused with premaxilla.
Character 2: Postrostral
    0, present; 1, absent.
Character 3: Supratemporal-intertemporal/dermopterotic area
    0, supratemporal larger than intertemporal; 1, intertemporal large, sutures with nasal; 2, bones fused or replaced by 'dermopterotic'.
    There is no intertemporal in *Platysiagum* and *Helmolepis* ('dermopterotic'), and this bone has probably fused with the supratemporal in *Acropholis stensioei*. Su (1993) identifies a single, comparatively small ossification in *Yuchoulepis*. Yet the presence of an intertemporal is unconfirmed.
Character 4: Outlines of frontals
    0, broad-short; 1, slender, oblong-rectangular.
    The skull roof is comparatively poorly known in some acrolepiforms, and it is obvious that with better preserved material of *Pygopterus*, *Acropholis*, and *Acrolepis*, their interrelationship will be better resolved.
Character 5: Length of maxilla
    0, shorter than dentary; 1, of same length as dentary.
Character 6: Postorbital plate of maxilla
    0, comparatively large, trapezoidal, and deep; 1, comparatively small, round, and narrow.

In *Ptycholepis marshi*, the maxilla is a tapering bar, and the character is therefore not applicable.

Character 7: Opercular region

0, operculum more oblique than suboperculum, oval-shaped, pointed and large; 1, intermediate; 2, operculum and suboperculum of about equal width.

Character 8: Accessory opercular bones (antoperculum)

0, none; 1, one comparatively large bone; 2, numerous small bones.

Plegmolepis kochi has a single large antoperculum plus numerous small ones; I decided to code for numerous ossifications rather than a single one (2).

Character 9: Counts of branchiostegal rays

0, numerous (up to 25); 1, about 10; 2, less than 10.

Character 10: Conspicuous enlargement of first branchiostegal ray

0, no; 1, yes.

Many taxa show a modified uppermost branchiostegal ray. *Platysiagum minus* has the comparatively smallest first ray (hardly enlarged if compared to other rays) and in *Acropholis stensioei*, the first branchiostegal ray has a peg that fits in the socket of the suboperculum's ventral margin (Aldinger, 1937: p. 46). In *Plegmolepis kochi*, the suboperculum issues a small ventral process in lieu of the slightly enlarged first branchiostegal ray.

Character 11: Gular region

0, lateral gulars larger than median gular; 1, lateral gulars absent or smaller than median gular.

Character 12: Supplementary rays of gill arches; *currently uninformative*

0, crescent-like; 1, needle-like.

This character is known in few taxa. However, it is unlikely to be a multistate character and should be easily coded in future analyses. The rays are typically crescent-like in *Hyllingea swanbergi* and needle-like in *Boreosomus arcticus*.

Character 13: Shape of preoperculum

0, slender-oblique; 1, dorsally expanded.

Character 14: Arrangement of postorbital ossifications

0, irregular pattern; 1, regular pattern; 2, none.

Character 15: Shape of infraorbital 3 (dermosphenotic)

0, comparatively large; 1, small and slender.

Character 16: Counts of extrascapulars

0, more than two pairs or numerous; 1, two pairs; 2, one pair.

Character 17: Presence of a presupracleithrum; *currently uninformative*

0, present; 1, absent.

The presence of a presupracleithrum may be an autapomorphy of *Mimia toombsi* but the state of this character is unknown in both *Watsonichthys* and *Hyllingea*, probably not closely related to *Mimia*.

Character 18: Dentition on oral jaw borders

0, numerous small teeth with fewer intercalated large teeth; 1, equal-sized teeth; 2, edentulous.

Character 19: Segmentation of lepidotrichia in pectoral fin
    0, entirely segmented; 1, distally segmented.
Character 20: Presence of spinous first ray in paired fins—*currently uninformative*
    0, no; 1, yes.
Character 21: Pelvic fin
    0, large with long basis; 1, small with short basis.
Character 22: Position of dorsal fin
    0, in posterior body half; 1, leading ray halfway between skull and caudal fin; 2, in anterior body half.
Character 23: Segmentation of lepidotrichia in dorsal fin—*currently uninformative*
    0, entirely segmented; 1, distally segmented.
    This character is currently unknown in five taxa; the character is included in this list for future reference.
Character 24: Anal fin with long base
    0, yes; 1, no.
Character 25: Segmentation of lepidotrichia in anal fin; *currently uninformative*
    0, entirely segmented; 1, distally segmented.
    This character is currently unknown in five taxa; the character is included in this list for future reference.
Character 26: Caudal fin shape
    0, hypercercal; 1, hypocercal; 2, nearly equilobate.
Character 27: Symmetry of caudal fin
    0, heterocercal; 1, hemiheterocercal.
Character 28: Basal fulcra in caudal fin
    0, yes; 1, no.

APPENDIX 6B: PTYCHOLEPID INTRARELATIONSHIPS— LIST OF CHARACTERS AND ANALYSIS

The following list of characters is modified from Wenz (1959, 1967), Schaeffer et al. (1975), Lund et al. (1995), and Mutter (2002).

Character 1: Postrostral (ch. 2 in Appendix 6A)
    0, present; 1, absent.
Character 2: Postorbital plate of maxilla (ch. 6 in Appendix 6A)
    0, comparatively large, trapezoidal, and deep; 1, comparatively small, triangular-trapezoidal; 2, comparatively small, round, and narrow.
Character 3: Shape of infraorbital 3 (dermosphenotic) (ch. 15 in Appendix 6A)
    0, comparatively large; 1, small and slender.
Character 4: Segmentation of lepidotrichia in pectoral fin (ch. 19 in Appendix 6A)
    0, entirely segmented; 1, distally segmented.

Character 5: Entopterygoid

0, several ossifications; 1, one ossification.

Character 6: Length ratio maxilla:dentary (modified from ch. 5 in Appendix 6A)

0, ratio = 1; 1, maxilla shorter than dentary.

Character 7: Shape of subopercular

0, quadrangular; 1, 'triangular' (posterior border twice as long as anterior border).

Character 8: Accessory opercular bones (antoperculum)

0, one comparatively large bone; 1, one comparatively small bone; 2, more than one bone.

Character 9: Counts of branchiostegal rays

0, numerous (up to 25); 1, about 10; 2, less than 10.

Character 10: Gular region

0, one large median gular; 1, two median gulars present; 2, lateral gulars present.

Character 11: Shape of preoperculum (modified from ch. 13 in Appendix 6A)

0, triradiate (dorsally moderately broad or slender); 1, triangular and inclined (dorsally broad); 2, only dorsally slender (ventral branch is absent).

Character 12: Number of postorbital ossifications (modified from ch. 14 in Appendix 6A)

0, many (7–15); 1, few (2–3).

Character 13: First branchiostegal ray (modified from former ch. 10 in Appendix 6A)

0, slightly or not enlarged; 1, enlarged.

ANALYSIS

See Figure 6.6 for consensus of trees. See Figures 6.7 through 6.12 for apomorphy list trees 1 through 6.

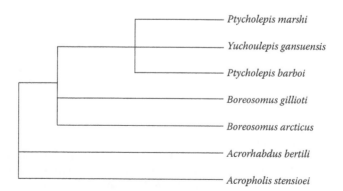

Ptycholepis marshi

Yuchoulepis gansuensis

Ptycholepis barboi

Boreosomus gillioti

Boreosomus arcticus

Acrorhabdus bertili

Acropholis stensioei

FIGURE 6.6 Consensus tree of Trees 1–6.

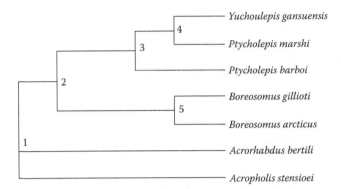

FIGURE 6.7 Apomorphy list tree 1: **Node 1→*Acropholis stensioei*** ch. 1(0→1); ch. 2(1→0); ch. 3(1→0); ch. 8(0→2); ch. 9(1→0). **Node 1→2** ch. 4(0→1); ch. 5(0→1); ch. 10(0→1); ch. 11 (1→0). **Node 2→3** ch. 6(0→1); ch. 7(0→1); ch. 8(0→1); 10(1→2); 13(0→1). **Node 3→*Ptycholepis barboi*** ch. 1(0→1); ch. 3(1→0). **Node 3→4** ch. 2(1→2). **Node 4→*Ptycholepis marshi*** ch. 11(0→2). **Node 4→*Yuchoulepis gansuensis*** ch. 7(1→0); ch. 12(0→1).

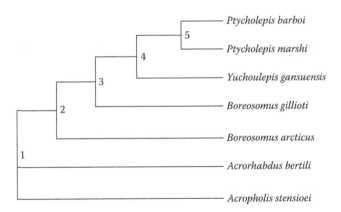

FIGURE 6.8 Apomorphy list tree 2: **Node 1→*Acropholis stensioei*** ch. 1(0→1); ch. 2(1→0); ch. 3(1→0); ch. 8(0→2); ch. 9(1→0). **Node 1→2**; ch. 4 (0→1); ch. 5(0→1); ch. 10(0→1); ch. 11(1→0). **Node 2→3** ch. 10(1→2). **Node 3→4** ch. 2(1→2); ch. 6(0→1); ch. 8(0→1); ch. 13(0→1). **Node 4→5** ch. 7(0→1). **Node 5→*Ptycholepis barboi*** ch. 1(0→1); ch. 2(2→1); ch. 3(1→0). **Node 5→*Ptycholepis marshi*** ch. 11(0→2). **Node 4→*Yuchoulepis gansuensis*** ch. 12(0→1).

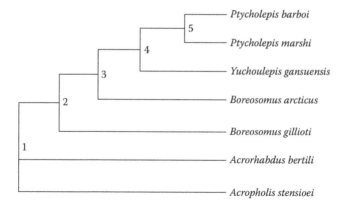

FIGURE 6.9 Apomorphy list tree 3: **Node 1→*Acropholis stensioei*** ch. 1(0→1); ch. 2(1→0); ch. 3(1→0); ch. 8(0→2); ch. 9(1→0). **Node 1→2** ch. 4(0→1); ch. 5(0→1); ch. 10(0→1); ch. 11(1→0). **Node 3→4** ch. 2(1→2); ch. 6(0→1); ch. 8(0→1); ch. 10(1→2); ch. 13(0→1). **Node 4→5** ch. 7(0→1). **Node 5→*Ptycholepis barboi*** ch. 1(0→1); ch. 2(2→1); ch. 3(1→0). **Node 5→*Ptycholepis marshi*** ch. 11(0→2). **Node 4→*Yuchoulepis gansuensis*** ch. 12(0→1).

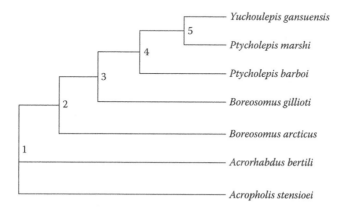

FIGURE 6.10 Apomorphy list tree 4: **Node 1→*Acropholis stensioei*** ch. 1(0→1); ch. 2(1→0); ch. 3(1→0); ch. 8(0→2); ch. 9(1→0). **Node 1→2** ch. 4(0→1); ch. 5(0→1); ch. 10(0→1); ch. 11(1→0). **Node 2→3** ch. 10(1→2). **Node 3→4** ch. 6(0→1); ch. 7(0→1); ch. 8(0→1); ch. 13(0→1). **Node 4→*Ptycholepis barboi*** ch. 1(0→1); ch. 3(1→0). **Node 4→5** ch. 2(1→2). **Node 5→*Ptycholepis marshi*** ch. 11(0→2). **Node 5→*Yuchoulepis gansuensis*** ch. 7(1→0); ch. 12(0→1).

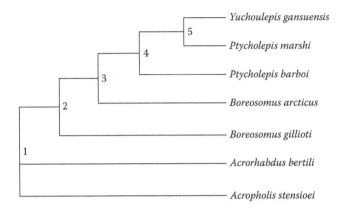

FIGURE 6.11   Apomorphy list tree 5: **Node 1→***Acropholis stensioei* ch. 1(0→1); ch. 2(1→0); ch. 3(1→0); ch. 8(0→2); ch. 9(1→0). **Node 1→2** ch. 4(0→1); ch. 5(0→1); ch. 10(0→1); ch. 11(1→0). **Node 3→4** ch. 6(0→1); ch. 7(0→1); ch. 8(0→1); ch. 10(1→2); ch.13(0→1). **Node 4→***Ptycholepis barboi* ch. 1(0→1); ch. 3(1→0). **Node 4→5** ch. 2(1→2). **Node 5→***Ptycholepis marshi* ch. 11(0→2). **Node 5→***Yuchoulepis gansuensis* ch. 7 (1→0); ch. 12(0→1).

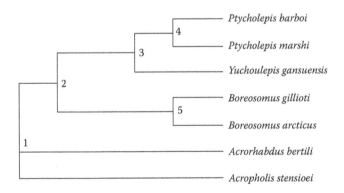

FIGURE 6.12   Apomorphy list tree 6: **Node 1→***Acropholis stensioei* ch. 1(0→1); ch. 2(1→0); ch. 3(1→0); ch. 8(0→2); ch. 9(1→0). **Node 1→2** ch. 4 (0→1); ch. 5(0→1); ch. 10(0→1); ch. 11(1→0). **Node 2→3** ch. 2(1→2); ch. 6(0→1); ch. 8(0→1); ch. 10(1→2); ch. 13(0→1). **Node 3→4** ch.7(0→1). **Node 4→***Ptycholepis barboi* ch. 1(0→1); ch. 2(2→1); ch. 3(1→0). **Node 4→***Ptycholepis marshi* ch. 11(0→2). **Node 3→***Yuchoulepis gansuensis* ch. 12(0→1).

LITERATURE CITED

Abel, O. (1906) Fossile Flugfische. *Jahrbuch der Kaiserlich-Königlichen Geologischen Reichsanstalt*, 56(1): 1–88, pls I–III.

Agassiz, L. (1832) Untersuchungen ueber die fossilen Fische der Lias-Formation. *Neues Jahbuch Mineralogie Geologie Paläontologie*, 1832: 139–149.

Agassiz, L. (1834) Abgerissene Bemerkungen über fossile Fische. *Neues Jahrbuch Geognosie etc.*, 1834: 379–390.

Agassiz, L.J.R. (1833) Neue Entdeckungen über fossile Fische. *Neues Jahrbuch Mineralogie, Geologie, Paläontologie*, 1833: 675–677.

Agassiz, L.J.R. (1844) *Recherches sur les poissons fossiles*. Petitpierre et Soleure, Neuchâtel, (1833–1844).

Aldinger, H. (1937) Permische Ganoidfische aus Ostgrönland. *Meddelelser om Grønland*, 102: 1–392, pls. 1–44.

Alessandri de, G.D. (1910) Studi sui pesci triasici della Lombardia. *Memorie della Società Italiana di Scienze Naturali e del Museo Civico di Storia Naturale di Milano*, 7(1): 1–145, 9 figs.

Algeo, T.J., Lehrmann, D., Orchard, M., and Jinnan, T. (eds.) (2007) The Permian-Triassic boundary crisis and Early Triassic biotic recovery. *Palaeogeography, Palaeoclimatology, Palaeoecology*, 252(1–2).

Andersson, E.A. (=Stensiö) (1916) Über einige Trias-Fische aus der Cava Trefontane, Tessin. *Bulletin of the Geological Institutions of the University of Uppsala*, 15: 13–33.

Andrews, S.M., Gardiner, B.G., Miles, R.S., and Patterson, C. (1967) Pisces. In *The Fossil Record* (eds.W.B. Harland, C.H. Holland, and M.R. House, et al.), Geological Society, London, pp. 637–683.

Bassani, F. (1886) Sui fossili e sull'età degli schisti bituminosi triasici di Besano in Lombardia. *Atti Societa Italiana Scienze naturali*, 29: 15–72.

Beltan, L. (1968) La fauna ichthyologique de l'Eotrias du N.W. de Madagascar: Le neurocrâne. *Cahiers de Paléontologie*, 1968: 1–135, pls I–V, I–L.

Beltan, L. (1984) Quelques poissons du Muschelkalk superieur d'Espagne. *Acta Geologica Hispanica*, 19: 117–127.

Beltan, L. (1996) Overview of systematics, paleobiology, and paleoecology of Triassic fishes of northwestern Madagascar. In *Mesozoic Fishes* (eds. G. Arratia and G. Viohl), Verlag Dr. Friedrich Pfeil, München, pp. 479–500.

Beltan, L.L. (1972) La faune ichthyologique du Muschelkalk de la Catalogne. *Memorias de la Real Academia de Ciencias y Artes de Barcelona*, 41: 281–325.

Benton, M.J. (ed.) (1993) *The Fossil Record 2*, Chapman & Hall, London.

Berg, L.S. (1958) *System der rezenten und fossilen Fischartigen und Fische*, VEB Deutscher Verlag der Wissenschaften, Berlin.

Bjerager, M., Seidler, L., Stemmerik, L., and Surlyk, F. (2006) Ammonoid stratigraphy and sedimentary evolution across the Permian-Triassic boundary in East Greenland. *Geological Magazine,* 143: 635–656.

Bronn, H.G. (1837) *Lethaea geognostica, oder Abbildungen und Beschreibungen der für die Gebirgs-Formationen bezeichnendsten Versteinerungen*, E. Schweizerbart, Stuttgart.

Bronn, H.G. (1858) Beiträge zur triasischen Fauna und Flora des bituminösen Schiefer von Raibl. *Neues Jahrbuch für Mineralogie, Geologie und Paläontologie*, 1: 1–32, 5 pls.

Brough, J. (1933) On a new Palaeoniscoid genus from Madagascar. *Annals and Magazine of Natural History* [series 10], 11: 76–87.

Brough, J. (1939) *The Triassic Fishes of Besano, Lombardy*, British Museum of Natural History, London.

Bürgin, T. (1992) Basal ray-finned fishes (Osteichthyes; Actinopterygii) from the Middle Triassic of Monte San Giorgio (Canton Tessin, Switzerland). *Schweizerische Paläontologische Gesellschaft*, 114: 1–164.

Chabakov, A. (1927) Synopsis of the ichthyofauna of the Permian deposits of Russia. *Zapiski Vserossijskogo Mineralogiceskogo Obscestva* [series 2], 56: 199–213.

Chang, M. and Miao, D. (2004) An overview of Mesozoic fishes in Asia. In *Mesozoic Fishes 3—Systematics, Paleoenvironments, and Biodiversity* (eds. G. Arratia and A. Tintori), Dr. Friedrich Pfeil, München, pp. 535–563.

Cloutier, R. and Arratia, G. (2004) Early diversification of actinopterygians. In *Recent Advances in the Origin and Early Radiation of Vertebrates* (eds. G. Arratia, M.V.H. Wilson, and R. Cloutier), Verlag Dr. Friedrich Pfeil, München, pp. 217–270.

Coates, M.I. (1999) Endocranial preservation of a Carboniferous actinopterygian from Lancashire, UK, and the interrelationships of primitive actinopterygians. *Philosophical Transactions of the Royal Society of London (series B)*, 354: 435–462.

Cope, E.D. (1887) Zittel's Manual of Palaeontology. *American Naturalist*, 21: 1014–1019.

Corroy, G. (1928) Les Vértebrés du Trias de Lorraine et le Trias Lorrain. *Annales de Paléontologie*, 17: 83–136.

Cotteau, G.-H. (1865) Note sur le *Ptycholepis bollensis* des calcaires bitumineux de Vassy (Yonne). *Bulletin de la Société Sciences Histoire Naturelles Yonne*, 19: 337–340.

Davis, J.W. (1884) Description of a new species of *Ptycholepis* from the Lias of Lyme Regis. *The Annals and Magazine of Natural History* [series 5], 13: 335–337, pl. 10.

Deecke, W. (1926) Animalia, Pars 33: Pisces triadici. In *Fossilium Catalogus I* (ed. C. Diener), W. Junk, Berlin, 201 pp.

Eastman, C.R. (1905) A brief general account of fossil fishes. *Geological survey of New Jersey (Annual Report of the State Geologist)*, Trenton 1904: 27–66.

Egerton, P. de M.G. (1852) British fossils [Descriptions of *Elasmodus, Palaeoniscus, Lepidotus, Pholidophorus, Ophiopsis, Ptycholepis, Leptolepis, Lophiostomus*]. *Memoirs of the Geological Survey of the United Kingdom* (Brit. organ. Rem.), 1852 [Dec. 6]: 10 articles.

Egerton, P. de M.G. (1854) On some genera and species of fossil fishes. *Annals and Magazine of Natural History London* [series 2], 13: 433–436.

Erwin, D.H. (2006) *How Life on Earth Nearly Ended 250 Million Years Ago*. Princeton University Press, Princeton, NJ.

Fitch, W.M. (1971) Toward defining the course of evolution: Minimum change for a specific tree topology. *Systematic Zoology*, 20: 406–416.

Gardiner, B.G. (1960) A revision of certain actinopterygian and coelacanth fishes, chiefly from the lower Lias. *Bulletin of the British Museum (Natural History)*, 4: 239–384, pls 36–43.

Gardiner, B.G. (1966) Catalogue of Canadian fossil fishes. *Life Sciences, Royal Ontario Museum, University of Toronto*, 68: 1–154.

Gardiner, B.G. (1967) Further notes on palaeoniscoid fishes with a classification of the Chondrostei. *Bulletin of the British Museum (Natural History)*, 14: 143–206.

Gardiner, B.G. (1984) The relationships of palaeoniscoid fishes, a review based on new specimens of *Mimia* and *Moythomasia* from the Upper Devonian of Western Australia. *Bulletin of the British Museum (Natural History), Geology*, 37(4): 173–428.

Gardiner, B.G. (1993) Osteichthyes: Basal Actinopterygians. In *The Fossil Record 2* (ed. M.J. Benton), Chapman & Hall, London, pp. 611–619.

Gardiner, B.G. and Schaeffer, B. (1989) Interrelationships of lower actinopterygian fishes. *Zoological Journal of the Linnean Society*, 97: 135–187.

Gardiner, B.G., Schaeffer, B., and Massarie, J.A. (2005) A review of the lower actinopterygian phylogeny. *Zoological Journal of the Linnean Society*, 144: 511–525.

Golonka, J. and Ford, D. (2000) Pangean (Late Carboniferous-Middle Jurassic) paleoenvironment and lithofacies, *Palaeogeography, Palaeoclimatology, Palaeoecology*, 161: 1–34.

Gross, W. (1935) Histologische Studien am Aussenskelett fossiler Agnathen und Fische. *Palaeontographica, Palaeozoologie*, Abt. A, 83: 1–60.

Hauff, B. (1960) *Das Holzmadenbuch*, Verlag der Hohenlohe'schen Buchhandlung, F. Rau, Öhringen.

Hay, O.P. (1902) Bibiliography and catalogue of the fossil Vertebrata of North America. *Bulletin of the United States Geological Survey*, 179: 1–868.

Hay, O.P. (1930) Second bibliography and catalogue of the fossil Vertebrata of North America. *Carnegie Institution Washington Publication*, 390 (2 vols): viii + 916 pp. (1929); xiv + 1074 pp.

Hennig, E. (1918) Über *Ptycholepis bollensis* Ag. *Jahreshefte Verein vaterländischer Naturkunde Württemberg*, 74: 173–182.

Hussakof, L. (1908) Catalogue of types and figured specimens of fossil vertebrates in the American Museum of Natural History. Part I - Fishes. *Bulletin of the American Museum of Natural History*, 25: 1–164.

Huxley, T.H. (1880) On the applications of the laws of evolution to the arrangement of the Vertebrata and more particularly of the Mammalia. *Proceedings of the Zoological Society of London*, 1880: 649–662.

Kner, R. (1866) Die Fische der bituminösen Schiefer von Raibl in Kärnthen. *Sitzungsberichte der kaiserlichen Akademie der Wissenschaften*, 53: 1–46.

Kner, R. (1867) Nachtrag zu den fossilen Fischen von Raibl. *Sitzungsbericht der kaiserlichen Akademie der Wissenschaften*, 5: 718–722, 1 pl.

Lamaud, P. (1977) Deux poissons holostéen du toarcien inférieur Franc-Comtois: *Ptycholepis bollensis* Agassiz et *Pachycormus curtus* Agassiz. *Bulletin Federal de la Société d'Histoire Naturelle Franc-Comté*, 78(1976/77): 37–49.

Lehman, J.-P. (1952) Etude complémentaire des poissons de l'Eotrias de Madagascar. *Kungliga Svenska Vetenskapsakademiens* [series 4], 2: 1–201.

Lehman, J.-P. (1966) Les Actinoptérygiens, Crossoptérygiens, Dipneustes. In *Traité de Paléontologie* (ed. J. Piveteau), Masson, Paris, pp. 1–387.

Liebermann, B.S. (1997) Early Cambrian paleogeography and tectonic history: A biogeographic approach. *Geology*, 25: 1039–1042.

Lieberman, B.S. (2000) *Paleobiogeography*. Kluwer Academic, New York.

Lieberman, B.S. and Eldredge, N. (1996) Trilobite biogeography in the Middle Devonian: Geological processes and analytical methods. *Paleobiology*, 22: 66–79.

Lund, R. (2000) The new actinopterygian order Guildayichthyiformes from the Lower Carboniferous of Montana (USA). *Geodiversitas*, 22: 171–206

Lund, R., Poplin, C., and McCarthy, K. (1995) Preliminary analysis of the interrelationships of some Paleozoic Actinopterygii. *Geobios*, 19: 215–220.

Morris, J. (1854) *A Catalogue of British Fossils. Comprising the Genera and Species hitherto Described; with References to their Geological Distribution, and to the Localities in which they have been Found.* John Morris, London.

Mutter, R.J. (2002) *Revision of the Triassic Family Colobodontidae (sensu Andersson 1916 (emended) with a Tentative Assessment of Perleidiform Interrelationships (Actinopterygii: Perleidiformes).* PhD Thesis, University of Zurich, 335pp.

Mutter, R.J. (2003) Reinvestigation of the Early Triassic ichthyofauna of the Sulphur Mountain formation (BC, Canada). *Canadian Paleontology Conference Proceedings*, 1: 32–36.

Mutter, R.J. (2005). Re-assessment of the genus *Helmolepis* Stensiö 1932 (Actinopterygii: Platysiagidae) and the Evolution of the Platysiagids in the Early-Middle Triassic. *Swiss Journal of Geosciences*, 98(2): 271–280.

Mutter, R.J. and Neuman, A.G. (2006) An enigmatic chondrichthyan with Paleozoic affinities from the Lower Triassic of western Canada. *Acta Palaeontologica Polonica*, 51: 171–182.

Mutter, R.J. and Neuman, A.G. (2008a) New eugeneodontid sharks from the Lower Triassic Sulphur Mountain Formation of western Canada. In *Fishes and the Break-up of Pangaea* (eds. L. Cavin, A. Longbottom, and M. Richter), Geological Society, London Special Publications 295: 9–41.

Mutter, R.J. and Neuman, A.G. (2008b) Jaws and dentition in an Early Triassic, 3-dimensionally preserved eugeneodontid skull (Chondrichthyes). *Acta Geologica Polonica*, 58: 223–227.

Mutter, R.J. and Neuman, A.G. (2009) Recovery from the end-Permian extinction event: Evidence from "Lilliput *Listracanthus*". *Palaeogeography, Palaeoclimatology, Palaeoecology*, 284: 22–28.

Neuman, A.G. and Mutter, R.J. (2005) *Helmolepis cyphognathus*, sp. nov., a new platysiagid actinopterygian from the Lower Triassic Sulphur Mountain Formation (British Columbia, Canada). *Canadian Journal of Earth Sciences*, 42: 25–36.

Newberry, J.S. (1878) XI. Descriptions of new fishes from the Trias. *Annals of the New York Academy of Sciences*, 1: 127–128.

Newberry, J.S. (1888) Fossil fishes and fossil plants. *Monographs of the United States Geological Survey*, 14: i–xiv, 1–152.

Nielsen, E. (1942) Studies on Triassic fishes from East Greenland –1. *Glaucolepis* and *Boreosomus*. *Paleozoologica Groenlandica*, 1: 1–394.

Nybelin, O. (1977) Studies on Triassic fishes from East Greenland III—on *Helmolepis gracilis* Stensiö. *Meddelelser om Grønland*, 200: 1–13.

Ørvig, T. (1978) Microstructure and growth of the dermal skeleton in fossil actinopterygian fishes: *Boreosomus, Plegmolepis,* and *Gyrolepis*. *Zoologica Scripta*, 7: 125–144.

Pitrat, C.W. (1973) Vertebrates and Permo-Triassic extinction. *Palaeogeography, Palaeoclimatology, Palaeoecology*, 14: 249–264.

Poplin, C. and Lund, R. (2000) Two new deep bodied palaeoniscoid actinopterygians from Bear Gulch (Montana, USA, Lower Carboniferous). *Journal of Vertebrate Paleontology*, 20: 428–449.

Priem, F. (1924) Paléontologie de Madagascar—XII.—Les poissons fossiles. *Annales de Paléontologie*, 13: 107–132.

Quenstedt, F.A. (1858) *Der Jura*. Laupp & Giebeck, Tübingen.

Quenstedt, F.A. (1867) *Handbuch der Petrefaktenkunde*. Verlag der Kaupp'schen Buchhandlung, Tübingen.

Rieppel, O. (1985) Die Triasfauna der Tessiner Kalkalpen XXV. Die Gattung *Saurichthys* (Pisces, Actinopterygii) aus der mittleren Trias des Monte San Giorgio, Kanton Tessin). *Schweizerische Paläontologische Abhandlungen*, 108: 1–103.

Sauvage, H.-É. (1875) Essai sur la faune ichthyologique de la périod liassique, suivi d'une notice sur les poissons fossiles du Lias de Vassy. *Annales des Sciences Géologiques* (Paris), 6: 1–58.

Sauvage, H.-É. (1891) Note sur quelques poissons du Lias supérieur de l'Yonne. *Bulletin de la Société des Sciences d'Histoire Naturelle, Yonne*, 45: 31–45, pls. 1–3.

Schaeffer, B. (1973) Fishes and the Permian-Triassic Boundary. In *The Permian and Triassic Systems and Their Mutual Boundary* (eds. A. Logan and L.V. Hills), Canadian Society of Petroleum Geologists Memoir 2.

Schaeffer, B. and Mangus, M. (1976) An early Triassic fish assemblage from British Columbia. *Bulletin of the American Museum of Natural History*, 156: 517–563.

Schaeffer, B., Dunkle, D.H., and McDonald, N.G. (1975) *Ptycholepis marshi* Newberry, a chondrostean fish from the Newark Group of eastern North America. *Fieldiana Geology*, 33: 205–233.

Schultze, H.-P. and Cumbaa, S.L. (2001) Chapter 18, *Dialipina* and the characters of basal actinopterygians. In *Major Events in Early Vertebrate Evolution. Palaeontology, Phylogeny, Genetics and Development* (ed. P.E. Ahlberg), Systematics Association, special volume series 61, Taylor & Francis, London and New York, pp. 315–332.

Schwarz, W. (1970) *Birgeria stensiöi* Aldinger. *Schweizerische Paläontologische Abhandlungen*, 89: 1–93.

Scupin, H. (1896) Vergleichende Studien zur Histologie der Ganoidschuppen. *Archiv für Naturgeschichte*, 62: 145–186, 2 pls.

Smith, A.B. (2007) Marine diversity through the Phanerozoic: Problems and prospects. *Journal of the Geological Society*, 164: 731–745.

Stemmerik, L., Bendix-Algreen, S.E., and Piasecki, S. (2001) The Permian-Triassic boundary in central East Greenland: Past and present views. *Bulletin of the Geological Society of Denmark*, 48: 159–167.

Stensiö, E.A. (1918) Notes on some fish remains collected at Hornsund by the Norwegian Spitzbergen expedition in 1917. *Norsk Geologisk Tidsskrift*, 5.

Stensiö, E.A. (1921) *Triassic Fishes from Spitzbergen*. Adolf Holzhausen, Wien.

Stensiö, E.A. (1932) Triassic fishes from East Greenland, collected by Danish expeditions in 1929–1931. *Meddelelser om Grønland*, 83: 1–305.

Su, D. (1993) New Jurassic ganoid fishes from northwestern Gansu, China. *Vertebrata PalAsiatica*, 31: 1–14.

Su, T.-T. (1974) New Jurassic ptycholepid fishes from Szechuan, S.W. China. *Vertebrata PalAsiatica*, 12: 1–15.

Swofford, D.L. (2000) *PAUP*. Phylogenetic Analysis Using Parsimony (*and Other Methods)*. Version 4, Sinauer, Sunderland, MA.

Sychevskaya, E.K. (2001) New data on the Jurassic freshwater fish fauna of northern Eurasia. In *Proceedings of the International Meeting on Mesozoic Fishes, Systematics, Paleoenvironments, and Biodiversity Abstracts*: 60 (ed. A. Tintori).

Véran, M. (1988) Les éléments accessories de l'arc hyoïdien des poissons téléstomes (Acanthodiens et Osteichthyens) fossiles et actuels. *Mémoires du Museum National d'Histoire Naturelle (Sciences de la Terre)*, 54: 1–98.

Wenz, S. (1959) Étude de *Ptycholepis bollensis*, poisson du Lias supérieur de l'Yonne et du Würtemberg. *Bulletin de la Société Géologique* [series 7], 1: 916–928.

Wenz, S. (1967) *Compléments a l'Étude des Poissons Actinoptérygians du Jurassique Français*, Editions C.N.R.S., Paris.

Williamson, W.C. (1849) On the microscopic structure of the scales and the dermal teeth of some ganoid and placoid fish. *Philosophical Transactions of the Royal Society of London*, 139: 435–475.

Woodward, A.S. (1891) *Catalogue of the Fossil Fishes in the British Museum (Natural History)*, part 2. British Museum (Natural History), London.

Woodward, A.S. (1895) *Catalogue of the Fossil Fishes in the British Museum (Natural History)*, part 3. British Museum (Natural History), London.

Woodward, A.S. (1896) On the fossil fishes of the Upper Lias of Whitby. Part I. *Proceedings of the Yorkshire Geological and Polytechnic Society*, 2: 25–42.

Woodward, A.S. (1912) Notes on some fish remains from the Lower Trias of Spitsbergen. *Bulletin of the Geological Institutions of the University of Uppsala*, 11: 291–297.

Xu, G.-H. and Gao, K.-Q. (2007) Early Triassic freshwater fishes from northern Gansu Province, China, and the age of the Beishan Beds. *Journal of Vertebrate Paleontology*, 27(3, suppl.): 169A.

Zigno, A. de (1891) Pesci fossili di Lumezzane in val Trompia. *Memorie della Reale Academia Lincei*, 4(7): 51–59.

Zittel, von K. A. (1890) *Handbuch der Paläontologie. Part 1. Paläozoologie, 3 - Vertebrata*, R. Oldenbourg, München and Leipzig.

7 Disparity as a Complement to Taxonomy and Phylogeny in Biogeographic Studies
Present and Past Examples from the Cephalopods

Alistair J. McGowan and Pascal Neige

CONTENTS

7.1   INTRODUCTION

Biogeography is a scientific discipline that is concerned with the geographic distribution of organisms across the earth (Upchurch et al., 2006). Thus it is essentially a question of spatial arrangement of biodiversity across a range of scales. At the broadest scale, biodiversity is spatially heterogeneous, but not randomly distributed. This nonrandom pattern has led to the suggestion of some 'law-like' mechanisms explaining observed distributional patterns. Well-known examples are latitudinal diversity gradients and Bergmann's law; the causal mechanism(s) involved in the generation of these patterns remains the topic of intense study in the modern biota (see Gaston and Blackburn, 2000 and references therein). Palaeontology, in common with other branches of the earth sciences, tends to adopt an approach of methodological uniformitarianism; preferring to invoke causes observable at the present time before seeking extraordinary explanations for observed evolutionary patterns. Jackson and Erwin (2006) discussed a number of studies designed to detect whether macroecological patterns can be identified in the fossil record. However, the evolutionary history of clades and their differential and more or less predictable responses to changes in the biological and physical environment are a major factor in the generation of biodiversity.

Explanations of the observed patterns and processes responsible for the generation and maintenance of biodiversity tend to be derived from measures of taxonomic diversity (e.g., species richness at various scales). Taxon counting has become a major focus in quantitative studies of biodiversity. Taxon counting has two major advantages over other measures of biodiversity. Species (or higher taxa in the case of many palaeontological studies) are relatively easy to define and count, and systematics is a common language across all clades, although there is some dispute about the equivalence of higher taxonomic ranks (Bertrand et al., 2006).

The limits of using taxonomy when studying biogeographic patterns are: (1) taxon counts do not fully quantify the differences amongst groups; (2) taxonomic richness is only one aspect of biodiversity. We are certainly not the first authors to argue the case for the use of multiple biodiversity metrics. Statements such as the following two have become increasingly common:

> Indeed, a large number of papers open with the recognition that species richness is only one measure of biodiversity but proceed to treat it as if it were *the* measure of biodiversity. (Gaston, 1996, p. 4)

> [T]axonomic patterns are only one aspect of biodiversity and may often mask interesting spatial patterns of morphological, functional, or phylogenetic diversity; or biodiversity may be quantified using a range of taxon-specific characteristics: species richness, genetic diversity, morphological disparity, body size and ecological niche breadth represent just a few possibilities. (Roy et al., 2004, p. 169)

This chapter is dedicated to the exploration of morphological disparity as a complement to taxonomic diversity when studying palaeobiogeographical or biogeographical patterns. Morphological disparity (hereafter disparity) has a number of desirable properties as a complementary biodiversity metric. Disparity captures the degree of difference between taxonomic units. Taxon counts do not provide any extra information about the amount of difference between two taxa, simply that there is enough divergence to establish two different taxa at a given rank. Disparity allows the identification of morphological traits that are tightly correlated with functional and ecological aspects of biodiversity that reflect adaptation to auto- and synecological aspects of the ecosystems to which organisms belong. Within the field of biogeographic/palaeobiogeographic studies, disparity has several advantages over other potential metrics, particularly genetic diversity and niche breadth that must be inferred to a greater or lesser degree. By using characters that can be identified in both extant and fossil taxa, disparity can provide a common denominator for bio- and palaeobiographic studies. Disparity has the potential to reveal spatial trends that differ from patterns revealed by taxon counts (Roy et al., 2001, 2004), thus providing richer insights into the dynamics underlying biogeographic patterns when compared with taxonomic metrics as this is part of the study of biodiversity through (geological) time (i.e., macroevolution).

As with any other measure of biodiversity, disparity has its own limits, as it can only capture some of the information about biodiversity, and should be used in conjunction with additional information and metrics. The most useful comparisons are restricted to organisms that share homologous features. However, Ricklefs and Miles (1994) demonstrated that it is possible to use Principal Component Analysis (PCA) of correlation matrices with the same number of characters to study organisms with different morphologies, although the exact characters may differ amongst the particular clades studied. As an example they compared the disparity amongst extant lizard and bird taxa within island communities to demonstrate the potential of this approach, reporting that birds had lower disparity than lizards.

The fossil record is primarily regarded as a record of evolutionary patterns through time. An unintended consequence of this view is that palaeontological studies tend to focus upon considering faunas and floras at only the global scale through time, while neglecting spatial analyses. This approach has been extremely fruitful in identifying temporal evolutionary patterns, with one of the best known being the three evolutionary faunas of Sepkoski (1981, 1993). Macroecological approaches to the fossil record have mostly focused on spatial distribution of taxonomic richness (see Jablonski et al., 2003 for an overview) or relationships between the longevity of taxa and their geographic ranges (Flessa and Jablonski, 1996; Miller, 1997; Hunt et al., 2005). Some ecologists have also taken a 'deep time' view of their data by projecting observed data and trends both forward and backward in time (Brown, 1995; Rosenzweig, 1995).

Resampling of morphological data has usually been used to assess differences between successive time intervals (e.g., Foote, 1992, 1993, 1996; Eble, 2000; Lockwood, 2004; Viller and Korn, 2004), however, it has also been used in a spatial context in the fossil record (McGowan, 2005; Navarro et al., 2005) to assess whether significant differences in disparity existed amongst regions to develop a spatiotemporal approach to the analysis of morphological evolutionary patterns in the fossil record.

Several recent morphometric studies have used techniques similar to those employed in the two studies presented here to investigate the spatial distribution of morphological variety amongst extant marine groups (Roy et al., 2001; Neige, 2003; McClain et al., 2004; McClain, 2005). Studies involving the fossil record have employed either single morphological measures such as body size/shell volume (Jablonski, 1997; Dommergues et al., 2002) or multivariate characterisations (Saunders and Swan, 1984; Foote, 1992, 1993, 1996; Wills et al., 1994; Roy, 1996; Wills 1998; Eble, 2000; Dommergues et al., 2001; Teusch et al., 2002; Navarro et al., 2005).

We present two studies: one on modern cuttlefish cephalopods (Neige) and one on ammonoids (McGowan), a group of extinct externally shelled cephalopods related to extant cephalopods such as cuttlefish and squid. Our objective is to demonstrate the potential of combining disparity and diversity metrics to gain a more complete view of patterns of biodiversity, and to examine the role that disparity information can play in formulating more detailed and refined hypotheses about evolutionary patterns and processes amongst extant and extinct clades.

7.2   RECENT SEPIID MORPHOLOGICAL EVOLUTION

We begin with an example of investigation of current geographical distribution patterns applied to the cuttlefish (Mollusca, Cephalopoda, Sepiidae), an extant group. Sepiidae is a speciose family with a wide variety of forms within the genera: *Sepia* Linnaeus 1758, *Sepiella* Gray 1849, and *Metasepia* Hoyle 1885. They possess a unique anatomical feature amongst living cephalopods (Figure 7.1): the dorsally embedded aragonitic shell known as the cuttlebone (or sepion) involved in buoyancy control (Denton and Gilpin-Brown, 1961). The possession of the cuttlebone is a key synapomorphy that defines Sepiidae as a monophyletic group.

Most sepiid cuttlefishes live in coastal waters. Although they are bottom dwellers cuttlefish are excellent swimmers when they leave the sea floor. Sepiid bathymetric distribution ranges from the sea surface down to 600 metres but sepiids are invariably

Post.   Ant.

FIGURE 7.1 A cuttlebone in ventral view (Sepia officinalis). Location of the 15 landmarks.

found close to the continental shelf or upper slope (Khromov, 1998). Sepiids are warm-water animals found in tropical, subtropical, and warm temperate regions of the Old World, but are completely absent from the Americas, the poles, and New Zealand. Only a few species occur in waters of the northern temperate zone. Their northern limits are Norway in the western part of their range and the Gulf of Peter the Great (Russia) in the eastern part. Sepiids were arranged in biogeographic units (Figure 7.2) mainly following Briggs (1995) and Khromov (1998). The latter used (species range) boundary compression as indicators of biogeographic boundaries. A total of 17 biogeographic units were established (labelled A–Q; see Figure 7.2). They spawn medium-sized eggs that are fixed to a substratum (Boletzky, 1998). No planktonic stages exist for young animals (Young et al., 1998). Thus, they are a model taxon for marine organisms lacking nondispersing eggs or young.

A detailed analysis of taxonomic diversity versus morphological disparity of the sepiids has been published (Neige, 2003). We propose here to sum up the results and to complement them with a recently published phylogenetic hypothesis based on genetic data (Bonnaud et al., 2006). As the diversity estimate from a genomic phylogeny is independent of morphology this may be regarded as an independent means of testing some of the statements made by Neige (2003) about how diversity/disparity ratios evolved in various regions.

7.2.1 Taxonomic Diversity in Space

From an exhaustive bibliographic analysis, 219 nominal sepiid species have been recorded. Only 111 are considered valid species here. This dataset is used for the present analysis and represents exhaustively what is known of sepiid taxonomy. Counts of species number by area reveal marked differences (Figure 7.2). Two main diversity 'hot-spots' appear at a global scale: the coast of southeast Africa (unit E), and a large area extending from eastern India to Japan and including Indonesia and the north coast of Australia (units K, L, M, and O). One important point is that species richness at northern boundaries is clearly different with just 4 species in unit A at the western end of the range versus 21 species in M in the east.

No apparent correlation between number of species and biogeographic unit size (compare, for example, units D and A) has been identified. Even if not formally tested, this seems to contradict the species/area effect (see Gaston and Blackburn, 2000). However, this relationship is not explored in detail here because unit size/area is difficult to quantify: cuttlefishes live near the bottom of the sea, so the biogeographic size unit measure would have to be based more on the effective available surface

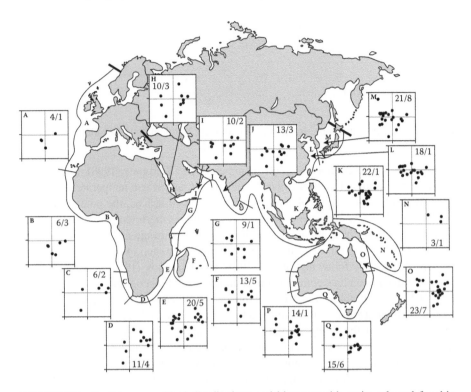

FIGURE 7.2   Sepiid geographical distributions and biogeographic units selected for this study. Each biogeographic unit is characterised by (1) a morphospace combining RW1 versus RW2 (see text) at the same scale, (2) species richness figured on the left, and (3) endemism figured on the right. A, from the northern limit at the western end of the sepiid range to the Tropic of Cancer (including Mediterranean Sea); B, from the Tropic of Cancer to latitude 20°S; C, from latitude 20°S to Cape Town; D, from Cape Town to Port Elizabeth; E, from Port Elizabeth to Zanzibar (along the African coast only); F, Madagascar and the Mascarene Ridge; G, from Zanzibar to Socotra; H, Red Sea; I, from Socotra to the Gulf of Oman (excluding Red Sea); J, from the Gulf of Oman to Sri Lanka, including the Arabian Gulf; K, from Sri Lanka (excluded) to the Gulf of Tonkin, including Malaysia, Indonesia and the Philippines; L, from the Gulf of Tonkin to Korea (west coast only); M, from Korea to the Gulf of Peter the Great, including Japan; N, New Guinea (north coast only), Melanesia and Micronesia; O, from the Tropic of Capricorn to Exmouth Gulf, including the south coast of New Guinea; P, from Exmouth Gulf to Perth; Q, from Perth to the Tropic of Capricorn, including Tasmania.

(thus taking into account the angle of slope of the continental shelf) than on simple 2D surface area. Sepiid diversity has no significant correlation with a range of environmental factors (e.g., salinity, sea-surface temperature, and windblown dust on the sea floor; see Skinner and Porter, 1999). However, several aspects of sepiid distribution merit closer attention. The northern boundary of the sepiid range is different at the western and eastern ends of their range at about 75°N and 45°N, respectively. This distribution seems to be linked to a summer sea-surface temperature barrier of about 12°C. A similar pattern is found for the southern boundary: 35°S and 12°C for South Africa and 40°S and 12°C for Tasmania. Diversity within these boundaries has no clear correlation with local temperature variations. Indeed the two diversity hot-spots have different sea-surface temperatures. The only exception, as previously noted by Khromov (1998), seems to be the relatively low diversity observed in unit C compared with equivalent latitudes off the eastern coast of Africa. The cool Benguela current, which hugs the southwest coast of Africa and is partly obstructed to the north by the Walvis Ridge, offers a physical explanation for this anomaly. Widely divergent salinity values occur within the geographic range of the sepiids (high for the Red Sea and Arabian Gulf, low for Indonesia) with no apparent correlation with species richness.

Cuttlefish global distributions can be studied in terms of latitude and longitude. The general pattern clearly shows that eastern geographic units typically have high species richness, whereas the western units have fewer species, except for unit E (Figure 7.2). The relationship amongst latitude, longitude, and species richness has been analysed statistically by linear regression (see Neige, 2003). No latitudinal or longitudinal gradients were demonstrated. The number of endemic species within a geographic unit is highly variable (Figure 7.2). Proportionally, it ranges from less than 5% (unit K) to 50% (unit B) with a mean value of 25%.

7.2.2 Morphological Disparity in Space

Landmark-based geometric morphometric techniques were applied in this study to the cuttlebones. The main feature of the landmark-based method is that it describes, quantifies, and localises geometrical differences amongst shapes regardless of size. Neige (2003) provides a detailed explanation of the morphometric procedures used. Fifteen points were chosen to describe the geometry of the different parts of the cuttlebones in ventral views (Figure 7.1). Note that cuttlebone shape is only one aspect of the cuttlefish morphology. Some other morphological features, such as tentacular club organisation, could provide some interesting complementary results but this study concentrates on the cuttlebone.

Relative warp analysis (Bookstein, 1991; Rohlf, 1994) was used to quantify the disparity pattern of cuttlebones, using the tpsRelw version 1.22 software (Rohlf, 1994) for analyses. A useful exploratory stage in morphometric work is to construct a 'morphospace' (i.e., a bi- or multivariate space into which taxa are ordinated as a function of their morphology). The first factorial plane (i.e., RW1 versus RW2), which generally concentrates a large amount of the information on morphological variance, illustrates the overall distribution pattern of shapes in the morphospace for the complete set of species or for a particular subset. In this case, the amount of total variation explained for the first five axes was $53 \times 0\%$, $20 \times 7\%$, $11 \times 5\%$, $5 \times 5\%$, and $4 \times 0\%$. Other axes are disregarded as they have eigenvalues of less than 1×0, indicating that they are not explaining any more of the variation than any single initial variable. The first relative warp describes the gross morphology of the sepion: positive values are returned for specimens with a broad sepion with large wings, and negative values for a narrow sepion with tiny wings. The second relative warp explains the shape of the inner cone, which may be long or short, regardless of general shape. The third relative warp focuses on the comparative length of the striated zone of the cuttlebone.

Co-ordinates of species in the multivariate morphospace can also be used as variables to assess the disparity of subsets of species (e.g., geographic groups). Parameters quantifying the distribution of a selected set of species of the morphospace have then to be computed. Three complementary disparity measures were used:

1. The amount of morphospace occupied by a given set of species is quantified using the sum of univariate ranges for the selected axes.
2. The position of a given set of species in the morphospace is quantified by the minimum and maximum values observed along the axes.
3. The average dissimilarity is given by the total variance, that is, the sum of variances for selected axes.

Rarefaction was used to eliminate the possible effect of differences in sample size on disparity estimates. Error bars on the three parameters describing disparity (sum of range, total variance, and position in morphospace) were calculated using a bootstrap process (500 runs). Rarefaction, bootstrap, and morphospace parameters were computed using the MDA package (Navarro, 2001). The three units with the highest total variance and range (C, D, and E) are those situated around southern Africa (Figure 7.3). All three units have range values that are greater than the standard deviation, and units C and D have total variance values that exceed the standard deviation. The lowest disparity values are for units A, B, and N. This may be because they are home to a small number of morphologically similar species. Total variance and range in other units lie between these extremes. Interestingly, unit K, corresponding partly to the 'East Indies' Triangle of Briggs (1999) and recognised as a species concentration zone (Briggs, 1999), has a relatively low disparity level compared with other units (Figure 7.3). This is mainly because of the presence of morphologically similar species. The minimum and maximum values on relative warps 1 to 3 indicate the position of the shapes for a given biogeographic unit in the global morphospace (Figure 7.3). This may be helpful in interpreting the fluctuations in the sum of univariate ranges. For example, minimum and maximum values on relative warps indicate that although units A and N have the smallest range values, these units display different patterns of morphospace occupation. The relationship amongst latitude, longitude, and disparity was analyzed using linear regression (see Neige, 2003). None of the tests were significant: no trends, either latitudinal or longitudinal, were found.

7.3 DISCUSSION

7.3.1 DISPARITY VERSUS DIVERSITY IN SPACE

No clear relationship exists between taxonomic diversity and disparity (Spearman's Rank Correlation Test), no matter which disparity metric is applied; the number of species in a given area does not predict the level of disparity. This emphasises the need for both diversity and disparity metrics to be explored when studying biogeography. Comparisons between diversity and disparity in the case of the cuttlefishes reveal various patterns. The most striking cases are

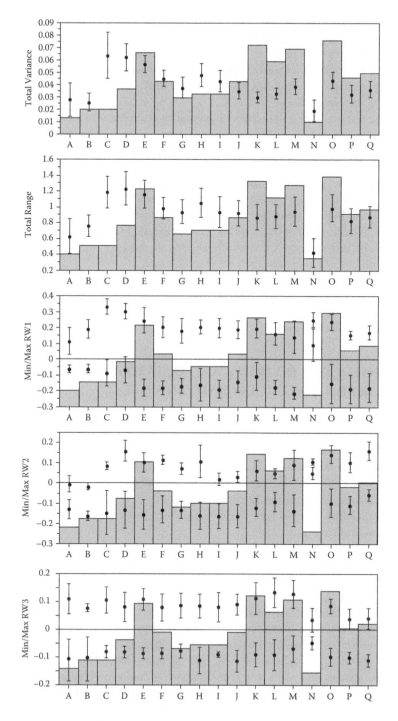

FIGURE 7.3 Disparity indices for the sepiids (labels refer to biogeographic units). Species richness is indicated by grey bars (no scale; see Figure 7.2 for exact values).

1. A high disparity around the coast of southern Africa, regardless of whether diversity (species richness) is high or low.
2. Relatively low disparity in the 'East Indies' area despite the high diversity (see Neige, 2003 for further results). The question of the biogeographic origins and history of sepiids may need to be explored in light of these results (and see Neige, 2003).
3. The southern African disparity high can be interpreted as the result of the co-existence of two independent phylogenetic clusters of species: one from the Atlantic Ocean and the other from the Indian Ocean. This distribution pattern could be explained using palaeogeographic data (Butterlin et al., 1993). The Tethys was still open at its eastern end during the Eocene and provided a connection between Europe's Mediterranean Sea on one side and the Indian Ocean and east African coasts on the other. At the end of the Eocene, this eastern corridor between the Mediterranean Sea and western India disappeared, completely altering the range of routes available for Sepiidae taxa to expand their geographic range.
4. The 'East Indies' area pattern (many species and low disparity) could indicate that the species in this area are all closely related, which explains the relative morphological similarity. Thus this area may be regarded as a centre of origin for Sepiidae.

7.4 PHYLOGENY, DIVERSITY, AND DISPARITY PROXIES FOR STUDYING BIODIVERSITY IN SPACE

We propose here some complementary explorations linking previous results (diversity, disparity, and diversity versus disparity) with phylogenetic ones. This has recently been made possible by the publication of a new phylogenetic framework for 19 sepiid species (Bonnaud et al., 2006) based on molecular data (Figure 7.4). Their final tree was obtained using simultaneous analysis of three genes: 12S rDNA, COII, and 16S rDNA. Amongst the 19 species they used, 17 are also used in the present diversity/disparity analysis (one was recently described and not used in the present study, and another was not identified at the species level in their paper). This allows some preliminary comparisons between disparity and phyletic patterns derived from molecular rather than morphological data. Molecular phylogenies provide an alternative dataset for estimating phylogenies independent of morphological characters (Roy et al., 2001). However, because of the markedly different numbers of species studied (111 for the disparity analysis, 19 for the molecular study), we must consider the following exploration as a preliminary one.

The molecular phylogeny (Bonnaud et al., 2006) may be divided into four monophyletic clusters, here labelled 1 to 4, and including respectively six, five, three, and three species also used in the disparity-based analysis (Figure 7.4). The biogeographic distribution of these four clusters is quite different. Three of them have nested patterns: strictly Australian and Malayan for Cluster 4; Australian, Malayan, Japanese, Indian, and northeast African for Cluster 3; and Australian, Malayan, Japanese, Indian, northeast African, and middle-east African for Cluster 1. Cluster 2

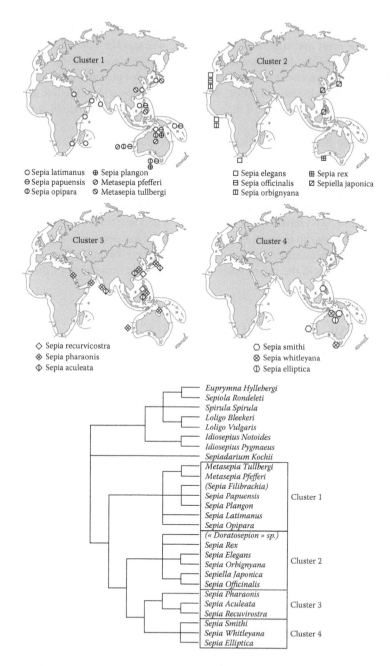

FIGURE 7.4 Biogeographic pattern for four phylogenetic patterns according to the phylogenetic relationships proposed by Bonnaud et al. (2006). Consensus tree (12S rDNA, COII, 16S rDNA). Clusters are determined in the present analysis (species not used in the disparity-based analysis [Neige, 2003] are in brackets). Species not assigned to any clusters are outgroups. (Modified from Bonnaud, L., Lu, C.C. and Boucher-Rodoni, R. (2006) *Biological Journal of the Linnean Society*, 89: 139–150. Figure 4, p. 146.)

displays a disjunct geographic pattern: some species in this cluster occur in Europe and along the west and south coasts of Africa, whereas the other members of this cluster occur along the Japanese, Malayan, and South African coasts. This disjunct distributional pattern could be explained by a number of possible factors: bias (sampling for phylogenetic analysis is not complete), migration and extinction, and vicariance (produced by the closure of the eastern end of the Tethyan ocean at the end of the Eocene as discussed above).

None of the three types of data used in this study (taxonomic diversity, morphologic disparity, and phylogenetic clustering) gives wholly congruent information about the biogeographical history of the sepiids. Thus it is worth exploring the distributional patterns of these animals in more detail. Roy et al. (2004) advocated the use of multiple datasets in their appeal to researchers to look 'beyond species richness'. Marquet et al. (2004) have argued in favour of 'a deconstruction of biodiversity patterns'. On this basis, an interesting analysis would be to compare morphological disparity to phylogenetic diversity (e.g., counting clusters) for any biogeographic unit. This could be done in the future when the phylogenetic framework of the sepiids is more complete.

7.5 TRIASSIC AMMONOID MORPHOLOGICAL EVOLUTION: A SPATIAL ANALYSIS

Ammonoids are a clade of extinct externally-shelled cephalopods which are superficially similar to living *Nautilus* but more closely related to coleoid cephalopods (squid, octopuses, and the cuttlefishes discussed by Neiger above; Engeser, 1996). Ammonoidea is a major monophyletic marine molluscan clade that ranges from the Early Devonian (Emsian 407–397 Ma) to the end-Cretaceous (65 Ma). The Triassic (251–199 Ma) is an intriguing interval in the evolutionary history of the group because all known genera, except for two, are apparently derived from a single common ancestor (Tozer, 1981a; Page, 1996), although some recent studies have challenged this scenario (Brayard et al., 2007; McGowan and Smith, 2007). As such, ammonoids appear to have considerable potential for rapid evolution, a feature sometimes termed 'evolvability' (Wagner and Altenberg, 1996; Kirschner and Gerhart, 1998; Hermida et al., 2002). Groups with the ability to rapidly generate significant diversity after biodiversity crises should be of interest to conservation biologists in the preparation of biodiversity action plans (BAPs).

Here we analyze the morphological evolution of Triassic ammonoids to search for evidence of spatial nonhomogeneity in disparity amongst the Triassic ammonoid marine biogeographic regions. Three major questions are explored in this study:

1. Did significant differences in disparity amongst regions exist during each stage and how do differences in richness within regions relate to disparity?
2. Are certain morphologies or morphological traits related to cosmopolitan or endemic distributions?
3. Within regions, are endemic taxa more or less disparate than a random subsample of all taxa known from within the region during that stage?

Answers to these questions provide a better understanding of the role of the spatial component of Triassic ammonoid morphological evolutionary patterns.

7.6 METHODS

7.6.1 DATA COLLECTION

Tozer (1981a) represents the most recent comprehensive revision of Triassic ammonoid genus-level taxonomy and has been the main authority for taxonomic decisions in this study. Tozer (1994) further revised Triassic ammonoids occurring within Canada, and the cosmopolitan distribution of many Triassic genera makes this monograph an important supplement to Tozer (1981a).

Morphological measurements were taken from specimens and published illustrations of 322 Triassic genera, with a single specimen from one species representing each genus. This sample represents ~62% of valid Triassic genera (Tozer, 1981a, 1994). The characters used in this study, listed in Figure 7.5 and Table 7.1, are a subset of those used by Saunders and Swan (1984). This character set combines the theoretical geometric coiling parameters of Raup (1966, 1967), which have been used in a number of studies of ammonoid morphological evolution, with other characters that record additional information about external shell morphology.

The Triassic Period was divided into eight time intervals (see the footnotes to Table 7.2 for interval names used in this study) to provide the temporal framework for this study. Some differences exist between these bins and the standard stages used in Gradstein et al. (2004). The Early Triassic Induan and Olenekian stages are subdivided into the Griesbachian and Dienerian substages and Smithian and Spathian substages, respectively. One continuing point of contention in Triassic stratigraphy is the status of the Rhaetian (see Krysten, 1990; Tozer, 1990 for reviews). Currently the Rhaetian is recognised as a stage and this status has become widely

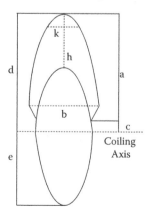

FIGURE 7.5 Diagram of hypothetical ammonoid with linear dimensions for calculating various morphological variables used to characterise ammonoid external shell shape. The formula for these calculations, along with additional variables, can be found in Table 7.1. (From McGowan, A.J., *Historical Biology*, 19:2 151-171: 2007. With permission from Taylor & Francis.)

TABLE 7.1

Characters Used to Describe Ammonoid External Shell Morphology in the Study[a]

Character (Abbreviation)	Derivation from Measurements
Whorl expansion (W)	$(d/e)^2$
Umbilical proportions (D)	c/d
Aperture shape (S)	b/a
Aperture height (AH)	f/a
Ventral structure	N/A
• Ventral groove < 1	
• Rounded = 0	
• Pointed > 1	
Ventral acuity (VW)	k/a (k = aperture width at a/10)
Ventro-lateral structure (VLG)	N/A
• Groove < 1	
• Rounded = 1	
• Pointed > 1	
Ribbing strength (RIB)	N/A
Relative strengths of radial and spiral ornament (TVS)	
• Spiral only = 0	
• Equally strong radial and spirial ornamentation = 0.5	
• Radial only = 1	
Measure of the continuity of tubercles between umbilicus and venter (LT)	Length of tubercle/flank width
Umbilical ridge strength (UR)	N/A
Number of constrictions on final whorl (CON)	N/A
Rib bifurcation (BIF)	Ratio of number of ribs at ventral edge to number at umbilical edge

Source: From McGowan, A.J., *Historical Biology*, 19:2 151-171: 2007. With permission from Taylor & Francis.

[a] The calculation column shows how the linear measurements shown in Figure 7.1 are used to calculate some of the variables.

TABLE 7.2

Disparity and Sample Size for Each Region Through the Triassic Period[a]

Regions	Interval[b]							
	G	D	Sm	Sp	A	L	C	N
Sample Size (Number of Genera)								
Arctic	10	20	25	10	17	12	13	10
West Pacific	—	14	34	12	15	7	11	5
East Pacific	—	26	27	24	41	38	44	58
Tethys	8	25	37	36	53	36	46	51
Germanic	—	—	—	1	5	3	—	—
Sephardic	—	—	—	1	5	3	—	—
Disparity								
Arctic	4.9	2.55	3.28 (*-)	6.27	5.59	6.53	6.86	6.4
West Pacific	—	2.98	3.93	5.36	5.53	9.91	5.28	5.54
East Pacific	—	2.94	4.44	6.11	5.6	6.58	6.01 (*-)	6.01
Tethys	04.36	2.98	4.04 (*-)	4.73 (*-)	5.47	7.24	7.11	6.05
Germanic	—	—	—	0	7.08	4.36	—	—
Sephardic	—	—	—	0	4.96	8.6	—	—

[a] Cases with lower observed disparity than expected, relative to 1,000 bootstrap replicates, are marked (*-).

[b] G = Griesbachian; D = Dienerian; Sm = Smithian; Sp = Spathian; A = Anisian; L = Ladinian; C = Carnian; N = Norian.

accepted (Gradstein et al., 1995; Golonka and Kiessling, 2002; Gradstein et al., 2004). However, due to the lack of clear global correlation, those parts of the Norian that are currently recognised as Rhaetian have been included in the Norian in the present analysis.

The stratigraphic ranges of many Triassic genera are summarised in Tozer (1981a). Ranges for genera described after this date were taken directly from the publication describing the genus. Tozer (1994) noted that ammonoid genus ranges at the stage level are probably overestimates. Given the extensive use of ammonoids for the correlation of rocks amongst regions (biostratigraphy), stage-level assignments are likely to be reliable amongst regions.

7.6.2 Spatial Distribution of Triassic Ammonoids

Several publications have summarised Triassic ammonoid palaeobiogeographic distributional patterns (Tozer, 1981a, 1981b; Dagys, 1988; Page, 1996). Ammonoids are not known from all provinces through all stages of the Triassic. The distribution of areas where Triassic ammonoids occur is summarised on a Triassic palaeogeographic map (Figure 7.6). The major biogeographic divisions used here are the result of faunal similarity analyses of Triassic ammonoid genera published by Kummel

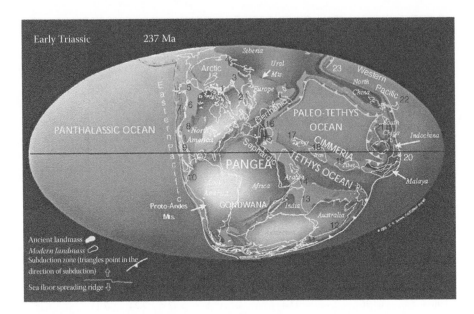

FIGURE 7.6 Mollewide palaeogeographic map (after Scotese PALEOMAP project [http://www.scotese.com/newpage8.htm]) of the Early Triassic world with ammonoid regions marked, and modern areas with Triassic ammonoid-bearing rocks numbered. Arctic Province (1. Canadian Arctic; 2. East Greenland; 3. Spitzbergen; 4. Siberia/Verkovnay). East Pacific (5. Alaska; 6. Western Canadian Basin; 7. Offshore terranes; 8. Western United States; 9. Mexico; 10. Peru/Columbia). Tethys (11. New Zealand; 12. Eastern Australia; 13. Kashmir/Pakistan; 14. Oman; 15. Balkans; 16. Alps; 17. Western Turkey; 18. Iran; 19. Tibet; 20. SE Asia; 21. South China). West Pacific (22. Japanese terranes; 23. South Primorye). The Germanic and Sephardic Provinces are marked with names only.

(1973) and Tozer (1981b). Other divisions are possible, such as the 20 areas used by Brayard et al. (2006) in a recent study of the spatial distribution of Early Triassic ammonoid diversity. Another possibility would be to use cladistic biogeographic techniques to delimit areas. However, the seven major units employed here provide a relatively fixed palaeobiogeographic framework throughout the Triassic that is suitable for the purposes of this analysis. The palaeobiogeographic divisions used in this study, and in previously published work on the Early Triassic biotic recovery of ammonoids after the End Permian mass extinction (McGowan, 2005), are from Page (1996). The paleobiogeographic units are listed below with additional information:

7.6.2.1 Tethyan Realm, Tethyan Subrealm

During the Early Mesozoic, the Tethyan Subrealm was a large ocean at low palaeolatitudes. Nearly 80% of all recognised Triassic ammonoid genera occurred in the Tethys and the Tethys has the highest proportion of endemic genera (~45%; Tozer, 1981b). Terranes repeatedly rifted from the northern margin of Gondwana, moving northwards before colliding with Asia. The western Tethys is a complex mosaic of micro-terranes; the closest modern analogue to the western Tethys would be the Indo-Malayan archipelago (Stampfli et al., 2001).

7.6.2.2   Tethyan Realm, Pacific Subrealm, East Pacific Province

Ammonoid-bearing rocks from the Eastern Pacific province outcrop along a long linear belt running north–south along what is now the west coast of the Americas, with some occurrences farther inland. Considerable variation exists in the palaeolatitudes of Triassic rocks from this region. The East Pacific Province has the second highest genus-level diversity. Despite a geographic separation of at least 150–180 degrees of longitude across Panthalassa, strong faunal similarities between the Tethyan and East Pacific faunas exist (Kummel, 1973; Tozer, 1981b; Page, 1996). However, understanding of ammonoid dispersal between the two provinces is confounded by the poor record of Triassic rocks in the Western Pacific province, which may conceal its importance as a stepping-stone in the exchange of ammonoid faunas between the East Pacific and Tethys from both the Arctic and East Pacific Regions.

A further complication is that many Triassic ammonoid faunas of the Eastern Pacific are recovered from complexes of accreted terranes on the western seaboard of North America (Tozer, 1981b, 1994). Although the exact means of emplacement and the distance of travel remain topics of active research (Trop et al., 2002), the presence of thick biogenic carbonates indicates that these terranes had their origins at lower latitudes (Jones et al., 1977). The terranes shown in Figure 7.6 (area 7) are just a few of the numerous terranes known from western North America (Tozer, 1994). Although elements of the Eastern Pacific faunas are recovered from these terranes, a considerable proportion of the fauna is recovered from rocks that had already accreted to the North American craton during the Triassic.

7.6.2.3   Tethyan Realm, Pacific Subrealm, West Pacific Province

The West Pacific ammonoid faunas are recovered from another series of accreted terranes representing rocks from deepwater facies. The record in this area is almost entirely based on chance accretion and, as a result, preservation of Triassic rocks is patchy. The ammonoid faunas from this region are a mixture of taxa known from the Arctic, Tethys, and East Pacific regions.

7.6.2.4   Boreal Subrealm, Arctic Province

The Triassic Arctic province had a similar configuration to its present day geography. This region is referred to as the 'Arctic', however, much of the province was at lower latitudes (<60°N) during the Triassic, and the North and South Poles were presumably ice-free and more temperate conditions would have existed (Wilson et al., 1994). The ammonoid-bearing rocks in this province include both siliciclastics and mixed carbonate clastic units. The Arctic had quite a diverse ammonoid fauna during the Early Triassic with many endemic taxa (Tozer, 1981b). This changes in the Middle and Late Triassic, when the sedimentary record becomes more influenced by fluvio-deltaic sediments (Embry, 1997), but ammonoids are known from the Arctic until the Norian.

7.6.2.5   Tethyan Realm, Germanic Province

Ammonoid-bearing rocks from the Germanic province are located in present-day Poland, Germany, and France. The Germanic province represents an area of epicontinental seas with depositional conditions cycling between marine and non-marine settings. Ammonoids are known from this area only from the Spathian–Early Carnian stages. The interchange of marine faunas between the Germanic and Tethyan provinces is tectonically controlled through three 'gates'. When these 'gates' closed, the Germanic province became virtually isolated (Urlichs and Mundlos, 1985). This is a similar situation to that described for the end-Eocene closure of the gateway at the eastern end of the Tethys in controlling the distribution of cuttlefish in the study presented above.

7.6.2.6 Tethyan Realm, Sephardic Province

The Sephardic Province was an ephemeral epicontinental seaway stretching from the Western Tethys to the mountains of present-day southern Spain. As with the Germanic province following the loss of connections with Tethys, a low-diversity endemic fauna developed during the Spathian–Ladinian (Tozer, 1981b).

The lack of ammonoid-bearing Triassic rocks in the modern Southern Hemisphere is evident from Figure 7.6, although the Triassic rocks now found in the Himalayas, Oman, and Australia were deposited south of the palaeoequator. Ammonoids from New Zealand represent the most southerly known palaeogeographic occurrences of Triassic ammonoids. Most Triassic rocks on the Gondwanan craton are either nonmarine or volcanic, and do not preserve marine fossils. In western Gondwana, Triassic ammonoid-bearing marine rocks of Spathian–Norian age are known from Mexico and from Middle and Late Triassic embayments in Peru, Columbia, and Chile, but no convincing evidence exists for a link between the Tethys and the East Pacific via a southern migratory pathway. Further work on the issue of whether temperature exhibited similar controls on ammonoids as it does on modern cuttlefish could be a promising area of research. Recent work by Brayard et al. (2006) provided global estimates of Early Triassic sea-surface temperatures using climate simulations that allowed them to identify latitudinal and longitudinal gradients amongst Early Triassic ammonoids.

7.7   DATA ANALYSIS

7.7.1   Disparity Analysis

The principal component (PC) I-III scores of the 322 Triassic genera derived from a PCA were used to assess whether regional subsets of taxa had higher or lower levels of disparity than would be expected if regional faunas were random subsets of the range of available morphologies within each stage. The sum of variances on the first three PC axes measures the observed level of disparity (Van Valen, 1974). Variance is a far better estimator of the underlying population than another widely-used morphospace occupation metric: the sum of ranges (Roy and Foote, 1997; Wills, 2001). To assess whether the observed disparity within each region was higher or lower than expected, the resampling scheme of Wills (1998) was used. Random subsamples of the same sample size were drawn from the sample of all genera known from a time interval and then compared with observed disparity. The Python 2.3 source code for this procedure is available upon request from AJM.

7.7.2 DISPARITY PATTERNS AMONGST COSMOPOLITAN AND ENDEMIC TAXA

Two end-member groups of taxa, cosmopolitan and endemic taxa, were distinguished based on their distribution amongst the regions. Cosmopolitan genera were operationally defined as genera recorded from all four major regions (Arctic, Tethys, East Pacific, and West Pacific). During the Griesbachian, ammonoids are only known from the Arctic and Tethyan regions, so any genus occurring in both regions during this interval was considered cosmopolitan. Endemic genera were defined as those recovered from only one region.

Ricklefs (2005) applied the following methodology to test whether species-poor passerine clades were preferentially concentrated at the edge of morphospace. Subsamples of appropriate size were repeatedly drawn from all taxa known from the stage. The mean PC score was calculated for this new sample. This procedure was used to test whether cosmopolitan and endemic taxa were significantly different in their mean PC scores, relative to the pool of all taxa within the region. By comparing the mean of the taxa identified as cosmopolitan and endemic with the distribution of 1,000 bootstrapped replicates, significant departures from expected values could be identified.

Wills' (1998) method was also employed to test whether endemic taxa within each region had a higher or lower disparity than a sample from the regional pool. The disparity of endemic taxa within the region was calculated and compared to the disparity of 1,000 bootstrap replicates drawn from all taxa found within that region during that stage.

7.8 RESULTS

7.8.1 PCA

The first three axes of the PCA analysis accounted for the following amounts of variation (PC I 18×0%; PC II 14×6%; PC III 12×5%; total variance accounted for ~45%). The morphological trends along each PC axis are as follows:

1. PC I: The morphological contrast between the two ends of this axis is between compressed (low-S) forms with acute venters, small umbilici, and weak ornamentation (few ribs and tubercles) at the negative end of the axis and more inflated (high-S) forms with larger umbilical areas (high-D) and stronger ornamentation at the positive end.
2. PC II: Taxa with high PC II scores tend to be compressed forms with high rates of whorl expansion (W), pronounced umbilical ridges, and bifurcating ribs. The negative end of this axis is typified by inflated forms with low rates of whorl expansion, constrictions, and ribs that tend to lack bifurcation.
3. PC III: Taxa with high PC III scores generally have strong ornament combined with low values of D, whereas forms with low PC III scores are weakly ornamented, with more open umbilici.

7.8.2 Disparity amongst Regions through Time

Table 7.2 shows the resampling results for regional disparity for each Triassic stage. In only 4 out of the 34 cases do regional faunas have disparity without the expected range. All four statistically significant cases record lower than expected disparity. Figure 7.7a plots disparity against sample size for all 34 cases. No significant correlation is observed.

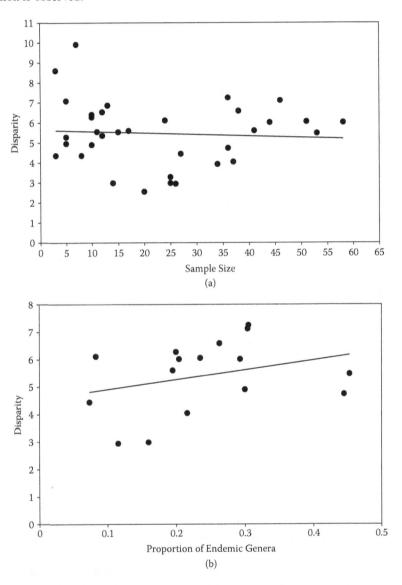

FIGURE 7.7 Biplot of the relationship of disparity to sample size and endemicity. (a) Genus richness versus disparity (Spearman Rank Correlation: $r = 0.30$; $n = 34$; N. S.) (b) Proportion of endemics versus disparity (Spearman Rank Correlation: $r = 0.07$; $n = 16$; N.S.).

TABLE 7.3

Number and Disparity of Endemic Genera Within Regions

Region	G	D	Sm	Sp	A	L	C	N
				Interval[a]				
			Sample Size (Number of Endemic Genera)					
Arctic	3	—	2	—	—	—	—	—
East Pacific	—	3	2	2	8	10	9	17
Tethys	—	4	8	16	24	11	14	12
			Disparity of Endemic Genera					
Arctic	9.11	—	2.2	—	—	—	—	—
East Pacific	—	8.41 (*+)	13.21 (*+)	34.20 (*+)	6.32	4.41	4.58	5.72
Tethys	—	5.33	4.82	5.47	5.33	6.73	8.66	5.22

[a] Higher than expected disparity marked (*+).

7.8.3 DISPARITY WITHIN REGIONS RELATIVE TO THE PROPORTION OF ENDEMIC GENERA

Table 7.3 shows the results of within-region comparison of endemic taxa relative to the other taxa known from within that region. The proportion of endemic taxa in a region was plotted against disparity to test the strength of any statistical correlation (Figure 7.7b). Although a statistically significant positive correlation exists, only about 10% of the observed variation is explained by variation in the number of endemic genera ($r^2 \sim 0.1$).

7.8.4 COMPARISON OF PC MEANS BETWEEN COSMOPOLITAN AND ENDEMIC GENERA

The mean scores of cosmopolitan genera (Figure 7.8a–c) and endemic genera (Figure 7.8f) on PCs I–III within each stage were compared with the distribution of 1,000 bootstrap replicates drawn from all taxa known from that particular stage. Amongst the pooled cosmopolitan genera there are three stages (Smithian, Ladinian, and Norian) during which the value of the PC I mean is significantly lower than that of the bootstrapped samples. No significant differences were found on the other two PC axes. Amongst the endemics there is never any significant difference between the observed mean for the endemic taxa within regions and random subsamples of the same size. Endemics form the largest portion of taxa in each region during most Middle and Late Triassic stages, and so greater confidence can be placed in these results than in those for the cosmopolitan genera. However, no single morphology, or suite of morphologies, is consistently associated with endemism.

Wilcoxon tests on the PC I–III scores of the cosmopolitan and endemic groups of taxa found no significant differences between the two groups (Figure 7.9a–c), which

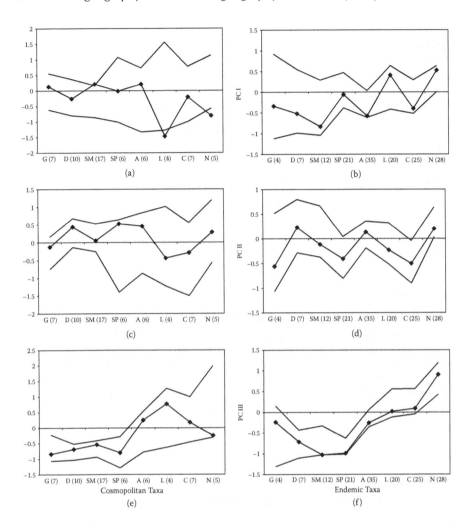

FIGURE 7.8 Plots of observed PC scores (diamonds) and the upper and lower 90% confidence intervals for the given sample size for each stage shown as dashed lines. (a–c) cosmopolitan taxa; (d–f) endemic taxa.

reinforces the point that no PC axis, and thus no combination of morphological traits, has significant predictive power as to which taxa are endemic and which are cosmopolitan.

The three PC axes considered here reduce a 13-dimensional space to three dimensions. It is possible that individual morphological characters may be significantly different between cosmopolitan and endemic taxa. Three characters were significantly different between the total pools of cosmopolitan and endemic genera. Cosmopolitan genera had significantly lower values of rib bifurcation ($\chi^2 = 8.93$; $p < 0.01$) and tubercle length ($\chi^2 = 4.04$; $p = 0.04$), and tended to have more rounded venters or ventral grooves ($\chi^2 = 4.88$; $p = 0.03$).

Body size (measured as shell diameter in millimeters) was significantly different between the cosmopolitan and endemic genera (Figure 7.9d; $\chi^2 = 17.26$; $p < 0.01$).

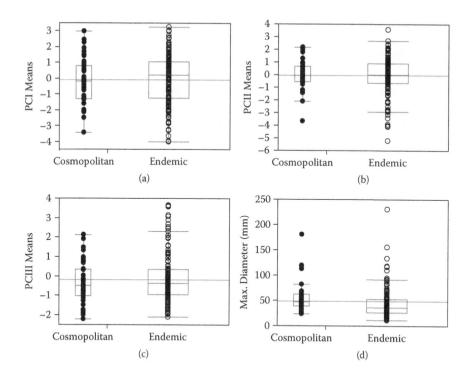

FIGURE 7.9 Wilcoxon tests for PC I-III scores (a–c) and body size (d) between pooled cosmopolitan ($n = 60$) and endemic ($n = 151$) taxa. (a) PC I ($\chi^2 = 0.55$; $p = 0.46$). (b) PC II ($\chi^2 = 0.18$; $p = 0.67$). (c) PC III ($\chi^2 = 1.2$; $p = 0.21$). (d) Body size (mm) ($\chi^2 = 17.26$; $p < 0.01$).

Cosmopolitan genera had a higher mean size (around 50 mm) than the endemic genera, although size varies considerably amongst the endemic genera.

7.8.5 Summary of Results

Amongst the major palaeobiogeographical regions considered by this study there was little evidence for the ammonoid faunas in any region having consistently higher or lower disparity throughout the Triassic (Tables 7.2 and 7.3). No statistically significant positive or negative correlation between genus-richness and disparity was found in analyses that considered firstly whole regional ammonoid faunas and then the proportion of endemic genera in a region (Figure 7.7). PCA axes defined by the study, which represent combinations of morphological traits, had no power to predict whether a particular genus would be cosmopolitan or endemic in its palaeobiogeographical distribution (Figure 7.9a–c), although some individual ornamentation characters were significantly different between cosmopolitan and endemic taxa. The final morphological trait to be considered was shell diameter as a proxy for body size. Cosmopolitan taxa did have a significantly higher mean shell diameter than endemic taxa, but considerable overlap exists between the two distributions (Figure 7.9d).

7.9   DISCUSSION

7.9.1   CAVEATS

Only ~62% of valid Triassic ammonoid genera were sampled. Some of the results reported above may change as more taxa are added into this sample. Another concern is that morphological data from a single specimen of a single species were used to represent all species of that genus in the analysis. However, given that species-level ammonoid taxonomy tends to focus on details of the suture line, this is probably not a serious problem.

Information on the spatial distribution of taxa is vulnerable to incomplete sampling (Koch, 1987). When moving to a regional level of interest, the question switches from 'Was this genus present, preserved, recovered, and recorded from somewhere in the world?' to 'Was this genus present, preserved, recovered, and recorded from this region?'. In regional analyses the vagaries of the fidelity of the fossil record, the range of ecosystems preserved within a region, and destruction of the rock record become ever more important factors. This ever-greater concern about sampling artefacts is a source of frustration. Attempts to analyse palaeobiological data in contexts that allow comparisons with evolutionary and ecological processes known from the present rely heavily on resolving the geographic scales at which we know these processes operate, and we must attempt to do so. Brayard et al. (2006) have argued that the basinal scale may be more appropriate for analyses of gradients, and this is certainly a future direction that we will have to consider.

Spatial data for this study were compiled from the primary and secondary literature. These data are vulnerable to biases based on the availability and publication date of the literature relating to different regions. The literature surveyed for the Triassic of North America is mainly from the 1950s onwards. This contrasts with the literature on the Tethyan fauna, which was published in the late nineteenth and early twentieth centuries. The quality of illustrations in the classic Tethyan monographs is excellent and these works remain essential sources of data on Triassic ammonoid morphology. However, the taxonomy of the classic monographs requires considerable revision. The almost complete lack of clear geological context of many Tethyan specimens was another problem. Indeed, much of the stratigraphy of Tethys has been reinterpreted since the realisation that the thin pelagic limestone beds found in Hallstatt-type facies were deposited over very long periods of time by processes with slow sedimentation rates (summarised in Tozer, 1981b).

Three relatively recent monographs (Silberling and Nichols, 1982; Bucher, 1988; Tozer, 1994) erected a number of new genera, mainly from the Middle and Late Triassic. Many of the genus-level diagnoses rely on quite small but specific details of ornamentation and shell shape, and the synonymy lists make it difficult to confidently assign older specimens from the Tethys to these new genera without inspection of the specimens themselves. This has probably led to overestimation of endemism in the East Pacific Province relative to the other regions.

7.9.2   DISPARITY ACROSS DIFFERENT SPATIAL SCALES

We know that biodiversity, measured as taxonomic diversity in the fossil record, is affected by sampling and spatial structure (Westrop and Adrain, 2001). Disparity needs to be examined in a similarly structured way if we are to better understand both modern and ancient ecosystems (McClain et al., 2004; McClain, 2005). This study of ammonoid disparity amongst Triassic regions indicates that genus-poor regions are not necessarily areas of low disparity; a finding shared by both Neige (2003 and above) and Roy et al. (2001). Plots of Triassic ammonoid disparity against sample size demonstrate that disparity is almost entirely uncorrelated with sample size, both within and amongst regions (Figure 7.7A,B; McGowan, 2005). By contrast, McClain (2005) found a positive relationship between diversity and disparity in deep-sea gastropods and argued this demonstrated that the environment controls disparity at sites only through limiting the number of species which can fit into a community, not through direct selection of particular morphologies.

Although genuine differences may exist in the number of taxa amongst regions used in this study, because of differences in the size of the regions, numbers of habitats, water masses, or latitudinal/longitudinal gradients such as those seen for present-day marine invertebrates, disparity shows no clear correlation with sample size. Likewise, the global low in disparity identified previously in McGowan (2004, 2005) during the Dienerian does not generate a regional disparity low within any of the individual regions.

Roy et al. (2001) demonstrated that close phylogenetic relatedness amongst taxa sampled offered one possible explanation of depressed levels of disparity, relative to taxonomic richness, amongst Indo-Pacific gastropods. If the endemic genera within a region are all from one or two families, this tends to result in a lower per-taxon increase in disparity due to close phylogenetic relatedness. Conversely, if all of the endemics in a region are from different families, this may lead to heightened disparity. Endemic taxa can arise by at least two pathways. They can either be a group that has relatives in another region or they can represent a localised diversification within a region. Kummel (1973) proposed that embayments in the western Tethys (Figure 7.6: Areas 15–17) were sources of elevated taxonomic diversity amongst Triassic ammonoids. Table 7.2 argues against this view, although further subdivision of the Tethys may identify more localised areas of elevated disparity. Table 7.3 shows that the only region with significant elevated disparity is the East Pacific Province, at small sample sizes (2 or 3 genera).

No simple relationship exists between genus-level endemism and disparity. During times of high endemism in the Tethys, disparity is lower than predicted. However, the East Pacific also has high levels of endemism during the Carnian and Norian. During the Carnian, disparity is lowered in the East Pacific rather than elevated. The overall positive correlation between the proportion of endemic genera and disparity is indicative that endemic taxa are exploring different regions of morphospace, rather than evolving similar forms, which would lead to a negative correlation between sample size and disparity. Other factors may well have been more important than endemism in the generation and regional structuring of Triassic ammonoid disparity.

Examination of the endemic Tethyan genera during the Smithian and Spathian shows that during the Smithian, although a number of families are present, nearly all are members of the superfamily Meekocerataceae. The genera are very similar in external morphology and are differentiated mainly on suture lines. During the Spathian the family Dinaritidae accounts for nearly all known Tethyan endemic taxa. The pattern in the Smithian of the Arctic remains anomalous. Although disparity is depressed within the Arctic fauna, the Arctic ammonoid genera sampled are from five superfamilies which is indicative of convergent or parallel evolution in this region.

This small sample of cases raises the question of which of the two mechanisms Roy et al. (2001) proposed for diversity/disparity mismatches is more likely to be responsible for observed Triassic mismatches at the regional level. One mechanism is close phylogenetic relationships amongst a large proportion of the taxa within a region. The other mechanism would be common environmental factors driving morphological convergence amongst distantly related taxa within each region. This stresses the need for ammonoid phylogenies, such as that of McGowan and Smith (2007), to differentiate between these two scenarios.

In future, framing detailed investigations of disparity and taxonomic richness, such as that of Klug et al. (2005), in a phylogenetic context within the Germanic and Sephardic provinces may represent an ideal way to further investigate this question at the level of local basins.

7.9.3 The Failure of Morphology to Predict Widespread and Endemic Genera

Some differences in the morphologies between cosmopolitan and endemic genera and overall sample pools were identified for individual stages. However, when the cosmopolitan and endemic genera were considered as single pooled samples, morphology had no clear ability to predict which genera were cosmopolitan and which endemic. The findings for the cosmopolitan genera were not as robust as those for the endemics because they were based on a small number of genera. Three of the individual morphological characters used to quantify morphology did show significant differences, and the general morphological suite of characters, including lack of ornament, high-W, low-S, and low-D, has often been associated with deepwater, cosmopolitan ammonoids (Jacobs et al., 1994; Jacobs and Chamberlain, 1996; Westermann, 1996).

The weak correlation between adult morphology and geographic range indicates that the mechanism(s) of transregional dispersal are probably not related to adult external shell morphology. This does not rule out a possible role for other aspects of adult shell morphology, such as the septa and siphuncular system which are so important for buoyancy regulation. Differences in life history traits, such as egg size or rate of growth, which adult morphology does not reflect may also be important (Westermann, 1996; Cecca, 2002).

Body size, measured as shell diameter, is the only trait significantly correlated with endemism. Measuring ammonoid 'size' as diameter is problematic because ammonoids with the same diameter can have quite different volumes depending on the width of the whorl section. Dommergues et al. (2002) considered size as volume, but even this approach of using an approximation based on a cylindrical model is vulnerable to differences in whorl shape. Nonetheless size (as diameter) appears to have some predictive power as to whether a genus is likely to be endemic or cosmopolitan.

It is tempting to argue that the significant observation is that above-average body size is the key predictor of how widespread a taxon is. However, a greater amount of variation in size is found amongst the endemic taxa, whereas the cosmopolitan taxa are quite closely clustered around their mean. Perhaps belonging to a particular size class, rather than merely being of above-average size, is the key to cosmopolitanism amongst Triassic ammonoids.

7.9.4 COMPARISON WITH PREVIOUS STUDIES OF AMMONOID PALAEOBIOGRAPHY AND DISPARITY

Dommergues et al. (2001) found little evidence of geographic controls on the spatial distribution of different Early Jurassic ammonoid morphotypes. Control was weakest during the earliest Jurassic. Thereafter only a few cases of certain morphologies corresponding to particular spatial distributions were identified. They found no consistency amongst time intervals as to which morphologies had specific distributions. One case in which Dommergues et al. (2001) did find a significant relationship between morphology and cosmopolitanism was amongst small platycones during the Domerian. These cosmopolitan forms were ~25 mm in diameter, about half the average size of cosmopolitan Triassic ammonoid genera.

The findings of this study are similar to the Early Jurassic results of Dommergues et al. (2001). No clear relationship amongst certain geographic distributions and certain morphologies exists. This suggests strong continuity in the processes that were shaping the regional biogeographical distributions of ammonoids amongstst regions during the Triassic and Jurassic. No clear evidence exists that those processes were sensitive to adult morphological traits.

Navarro et al. (2005) studied the effect of ammonite faunal invasions in the Boreal Realm during the Jurassic. Faunal invasions tended to depress disparity among the 'native' ammonoid fauna, and they noted that change occurs in terms of the areas of morphospace that clades occupy rather than drastic reorganisations of morphospace. Origination and extinction dynamics were important in generating changes in the disparity of individual clades, but were less important for extending the range or variety of morphologies among the Boreal Jurassic ammonites. Once again the relationship between spatial distribution and disparity was ambiguous.

7.10 CONCLUSIONS

No clear pattern of systematic regional differences in morphological variation existed amongst Triassic ammonoid genera. Genuine differences in richness may exist amongst regions, however, there is no evidence that these differences in richness are related to either lesser or greater than expected levels of disparity. Most departures from expected levels of disparity are toward significantly lower levels, which may indicate some role for endemism in driving such patterns, but a general analysis shows a small but significant positive correlation between the proportion of endemics in a region and disparity.

No clear link has been established, at either the stage level or for the whole of the Triassic, between the adult external shell morphology of Triassic ammonoids and their palaeobiogeographical distribution patterns. Cosmopolitanism appears to be the product of random dispersal rather than an emergent property of certain adult morphotypes.

7.11 CONCLUDING THOUGHTS AND FUTURE DIRECTIONS

We have presented two studies highlighting the potential of disparity as a complementary biodiversity metric to diversity. A common theme that emerges from these studies of fossil and extant cephalopods is that taxonomic diversity alone gives an incomplete measure of biodiversity. Taxonomic diversity does not have strong predictive power about disparity either at global scales or within individual biogeographic units. This may appear to be counterintuitive as logically some disparity must be added by every new taxon sampled, particularly in palaeontological studies where taxa are usually defined on morphological differences alone. However, all that the erection of new taxa indicates is a minimal disparity threshold has been crossed on which to base a new taxon. Any additional disparity beyond that, but not sufficient to warrant the establishment of a new taxon at the next taxonomic level, will add to disparity without adding to taxonomic diversity.

The analytical tools used in the two case studies are widely available to carry out morphometric and disparity analyses. A list of morphometric software for data collection and analysis can be found at www.palass.org under the links section. PAST (Hammer et al., 2001) is a freeware package which, although aimed at palaeontologists, is ideal for all evolutionary biologists and provides an excellent way to get started in morphological and disparity analyses.

Penny and Phillips (2004) made a strong case for developing biologically-realistic models that combined more data types, in the context of understanding diversification after biodiversity crises. The notion that we should be combining phylogenetic, morphological, and ecological data into our analyses to build more rigorous models should be applied much more widely in evolutionary biology. Morphology is often a useful proxy for ecological data and many more sources of data are available for ecological variables in recent systems, as the study of cuttlefish presented here emphasises. Phylogenies, particularly molecular phylogenies, are the other aspect that can play a major role in deciding amongst processes responsible for diversity/disparity patterns, particularly mismatches.

The other issue that needs to be addressed is how disparity is 'packed' into the biosphere. We only have a handful of studies on disparity packing, most of which are at reasonably large spatial scales, and then the contrasting results of McClain (2005) which suggest that local disparity is a subsample of regional disparity. Studies by Sepkoski (1988) and Westrop and Adrain (2001) addressed in great detail the issue of how palaeobiodiversity is structured at alpha, beta, and gamma levels, and gained insight into the relationship amongst the levels and how new diversity was added to the biosphere through time. Disparity studies using a similar set of nested hierarchies for analyses could be a powerful tool for understanding how additional disparity is generated and spatially distributed in the fossil record and extant ecosystems. We hope this contribution will stimulate interest in disparity analyses amongst biogeographers working on both extinct and extant taxa.

ACKNOWLEDGEMENTS

This study is a contribution to the Team Forme, Evolution, Diversité of the UMR CNRS5561, and to the GDR 2474 Morphométrie et Evolution des Formes. Much of the section on regional analysis of Triassic ammonoid disparity was developed during AJM's PhD research at the University of Chicago. AJM thanks his PhD committee D. Jablonski, M. Foote, S. M. Kidwell, W. B. Saunders, and A. M. Ziegler. The manuscript, and additional analytical work, was completed at the Natural History Museum, London, with funding from the Leverhulme Trust to A. B. Smith.

LITERATURE CITED

Bertrand, Y., Pleijel, F., and Rouse, G.W. (2006) Taxonomic surrogacy in biodiversity assessments, and the meaning of Linnaean ranks. *Systematics and Biodiversity*, 4: 149–159.

Boletzky, S.V. (1998) Cephalopod eggs and egg masses. *Oceanography and Marine Biology: An Annual Review*, 36: 341–371.

Bonnaud, L., Lu, C.C., and Boucher-Rodoni, R. (2006) Morphological character evolution and molecular trees in sepiids (Mollusca: Cephalopoda): Is the cuttlebone a robust phylogenetic marker? *Biological Journal of the Linnean Society*, 89: 139–150.

Bookstein F.L. (1991) *Morphometric Tools for Landmark Data, Geometry, and Biology*, Cambridge University Press, Cambridge, UK.

Brayard, A., Bucher, H., Bruhwiler, T., Galfetti, T., Goudemand, N., Guodun, K., Escarguel, G., and Jenks, J. (2007) *Proharpoceras Chao*: A new ammonoid lineage surviving the end-Permian mass extinction. *Lethaia*, 40: 175–181.

Brayard, A., Bucher, H., Escarguel, G., Fluteau, F., Bourquin, S., and Galfetti, T. (2006) The Early Triassic ammonoid recovery: Paleoclimatic significance of diversity gradients. *Palaeogeography, Palaeoclimatology, Palaeoecology*, 239: 374–395.

Briggs, J.C. (1995) *Global Biogeography*, Elsevier, Amsterdam.

Briggs, J.C. (1999) Coincident biogeographic patterns: Indo-West Pacific ocean. *Evolution*, 53: 326–335.

Brown, J.H. (1995) *Macroecology*, University of Chicago Press, Chicago.

Bucher, H. (1988) A new Anisian (Middle Triassic) ammonoid zone from northwestern Nevada. *Eclogae Geologicae Helvetiae*, 81: 723–762.

Butterlin, J., Vrielynck, B., Bignot, G., Clermonte, J., Colchen, M., Dercourt, J., Guiraud, R., Poisson, A., and Ricou, L.E. (1993) Lutetian (46 to 40 Ma). In *Atlas of Tethys Palaeoenvironmental Maps. Explanatory Notes* (eds. J. Dercourt, L.E. Ricou, and B. Vrielynck), Gauthier-Villars, Paris, pp. 197–209.

Cecca, F. (2002) *Palaeobiogeography of Marine Fossil Invertebrates - Concepts and Methods*, Taylor and Francis, London.

Dagys, A.S. (1988) Major features of the geographic differentiation of Triassic ammonoids, In *Cephalopods Present and Past* (eds. J. Wiedmann and J. Kullmann), Schweizerbart'sche Verlangbuchhandlung, Stuttgart, pp. 341–349.

Denton, E.J. and Gilpin-Brown, J.B. (1961) The buoyancy of the cuttlefish, *Sepia officinalis* (L.). *Journal of the Marine Biological Association*, 41: 319–342.

Dommergues, J.-L., Laurin, B., and Meister, C. (2001) The recovery and radiation of Early Jurassic ammonoids: Morphologic versus palaeobiogeographic patterns. *Palaeogeography, Palaeoclimatology, Palaeoecology*, 165: 195–213.

Dommergues, J.-L., Montuire, S., and Neige, P. (2002) Size patterns through time: The case of the Early Jurassic ammonite radiation. *Paleobiology*, 28: 423–434.

Eble, G.J. (2000) Contrasting evolutionary flexibility in sister groups: Disparity and diversity in Mesozoic atelostomate echinoids. *Paleobiology*, 26: 56–79.

Embry, A.F. (1997) Global sequence boundaries of the Triassic and their identification in the Western Canada sedimentary basin. *Bulletin of Canadian Petroleum Geology*, 45: 415–433.

Engeser, T. (1996) The position of the Ammonoidea within the Cephalopoda. In *Ammonoid Paleobiology* (eds. N.H. Landman, K. Tanabe, and R.A. Davis), Plenum, New York, pp. 3–22.

Flessa, K.W. and Jablonski, D. (1996) The geography of evolutionary turnover: A global analysis of extant bivalves. In *Evolutionary Paleobiology* (eds. D. Jablonski, D.H. Erwin, and J.H. Lipps), University of Chicago Press, Chicago, pp. 376–397.

Foote, M. (1992) Rarefaction analysis of morphological and taxonomic diversity. *Paleobiology*, 18: 1–16.

Foote, M. (1993) Discordance and concordance between morphological and taxonomic diversity. *Paleobiology*, 19: 185–204.

Foote, M. (1996) Ecological controls on the evolutionary recovery of post-Paleozoic crinoids. *Science*, 274: 1492–1495.

Gaston, K.J. (1996) What is biodiversity? In *Biodiversity, a Biology of Numbers and Difference* (ed. K.J. Gaston), Blackwell Science, Oxford, pp. 1–9.

Gaston, K.J. and Blackburn, T.M. (2000) *Pattern and Process in Macroecology*, Blackwell Science, Oxford.

Golonka, J. and Kiessling, W. (2002) Phanerozoic time scale and definition of time slices. *Phanerozoic Reef Patterns. SEPM Special Publication*, 72: 11–20.

Gradstein, F.M., Agterberg, F.P., Ogg, J.G., Hardenbol, J., Van Veen, P., Thierry, J., and Huang, Z. (1995) A Triassic, Jurassic, and Cretaceous time scale. *Geochronology, Time Scale, and Global Stratigraphic Correlation SEPM Special Publication*, 54: 95–121.

Gradstein, F.M., Ogg, J.G., Smith, A.G., Bleeker, W., and Lourens, L.J. (2004) A new geologic time scale with special reference to Precambrian and Neogene. *Episodes*, 27: 83–100.

Hammer, Ø., Harper, D.A.T., and Ryan, P.D. (2001). Past: Paleontological Statistics Software Package for Education and Data Analysis. *Palaeontologia Electronica*, 4(1), art. 4: 9pp., 178kb. http://palaeo-electronica.org/2001_1/past/issue1_01.htm.

Hermida, M., Fernández, C., Amaro, R., and San Miguel, E. (2002) Heritability and "evolvability" of meristic characters in a natural population of *Gasterosteus aculeatus*. *Canadian Journal of Zoology*, 80: 532–541.

Hunt, G., Roy, K., and Jablonski, D. (2005) Species-level heritability reaffirmed: A comment on 'On the heritability of geographic range sizes'. *The American Naturalist*, 168: 129–135.

Jablonski, D. (1997) Body-size evolution in Cretaceous molluscs and the status of Cope's rule. *Nature*, 385: 250–252.

Jablonski, D., Roy, K., and Valentine, J.W. (2003) Evolutionary macroecology and the fossil record. In *Macroecology: Concepts and Consequences* (eds. T.M. Blackburn and K.J. Gaston), Blackwell Scientific, Oxford, pp. 368–390.

Jackson, J.B.C. and Erwin, D.H. (2006) What can we learn about ecology and evolution from the fossil record? *Trends in Ecology and Evolution*, 21: 322–328.

Jacobs, D.K. and Chamberlain, J.A., Jr. (1996) Buoyancy and hydrodynamics in ammonoids. In *Ammonoid Paleobiology* (eds. N.H. Landman, K. Tanabe, and R.A. Davis), Plenum, New York, pp. 169–251.

Jacobs, D.K., Landman, N.H., and Chamberlain, J.A., Jr. (1994) Ammonite shell shape covaries with facies and hydrodynamics: Iterative evolution as a response to changes in basinal environment. *Geology*, 22: 905–908.

Jones, D.S., Silberling, N.J., and Hillhouse, J. (1977) Wrangellia—A displaced terrane in north-western North America. *Canadian Journal of Earth Sciences*, 14: 2565–2577.

Khromov, D.N. (1998) Distribution patterns of Sepiidae. *Smithsonian Contributions to Zoology*, 586: 191–206.

Kirschner, M. and Gerhart, J. (1998) Evolvability. *Proceedings of the National Academy of Sciences, USA*, 95: 8420–8427.

Klug, C., Schatz, W., Korn, D., and Reisdorf, A.G. (2005) Morphological fluctuations of ammonoid assemblages from the Muschelkalk (Middle Triassic) of the Germanic Basin—Indicators of their ecology, extinctions, and immigrations. *Palaeogeography, Palaeoclimatology, Palaeoecology*, 221: 7–34.

Koch, C.F. (1987) Prediction of sample size effects on the measured temporal and geographic distribution patterns of species. *Paleobiology*, 13: 100–107.

Krysten, L. (1990) A Rhaetian stage—Chronostratigraphy, subdivisions, and their intercontinental correlation, *Albertina*, 8: 15–24.

Kummel, B. (1973) Lower Triassic (Scythian) mollusks. In *Atlas of Palaeobiogeography* (ed. A. Hallam), Elsevier, Amsterdam, pp. 225–233.

Lockwood, R. (2004) The K/T event and infaunality: Morphological and ecological patterns of extinction and recovery in veneroid bivalves. *Paleobiology*, 30: 507–521.

Marquet, P.A., Fernandez, M., Navarrete, S.A., and Valdovinos, C. (2004) Diversity emerging: Toward a deconstruction of biodiversity patterns. In *Frontiers of Biogeography. New Directions in the Geography of Nature* (eds. M.V. Lomolino and L.R. Heaney), Sinauer, Sunderland, MA, pp. 191–209.

McClain, C.R. (2005) Bathymetric patterns of morphological disparity in deep-sea gastropods from the Western North Atlantic Basin. *Evolution*, 59: 1492–1499.

McClain, C.R., Johnson, N.J., and Rex, M.A. (2004) Morphological disparity as a biodiversity metric in lower bathyal and abyssal gastropod assemblages. *Evolution*, 58: 338–348

McGowan, A.J. (2004) Ammonoid taxonomic and morphologic recovery patterns after the Permian-Triassic. *Geology*, 32: 665–668.

McGowan, A.J. (2005) Ammonoid recovery from the Late Permian mass extinction event. *Comptes Rendus Palevol*, 4: 449–462.

McGowan, A.J. (2007) Does shape matter? Morphological diversity and differential survivorship among Triassic ammonoid genera, *Historical Biology*, 19: 151–171.

McGowan, A.J. and Smith, A.B. (2007) Ammonoids across the Permian/Triassic boundary: A cladistic perspective. *Palaeonotology* 50: 573–590.

Miller, A.I. (1997) Dissecting global diversity patterns: Examples from the Ordovician radiation. *Annual Review of Ecology and Systematics*, 28: 85–104.

Navarro, N. (2001) *MDA: Morphospace-Disparity Analysis for Matlab, Version 1.1.* Paléontologie Analytique, Dijon.

Navarro, N., Neige, P., and Marchand, D. (2005) Faunal invasions as a source of morphological constraints and innovations? The diversification of the early Cardioceratidae (Ammonoidea; Middle Jurassic). *Paleobiology*, 31: 98–116.

Neige, P. (2003) Spatial patterns of disparity and diversity of the Recent cuttlefishes (Cephalopoda) across the Old World. *Journal of Biogeography*, 30: 1125–1137.

Page, K.N. (1996) Mesozoic ammonoids in time and space. In *Ammonoid Paleobiology* (eds. N.H. Landman, K. Tanabe, and R.A. Davis), Plenum, New York, pp. 755–794.

Penny, D. and Phillips, M.J. (2004) The rise of birds and mammals: Are microevolutionary processes sufficient for macroevolution? *Trends in Ecology and Evolution*, 19: 516–522.

Raup, D.M. (1966) Geometric analysis of shell coiling: General problems. *Journal of Paleontology*, 40: 1178–1190.

Raup, D.M. (1967) Geometric analysis of shell coiling: Coiling in ammonoids. *Journal of Paleontology*, 41: 43–65.

Ricklefs, R.E. (2005) Small clades at the periphery of passerine morphological space. *The American Naturalist*, 165: 651–659.

Ricklefs, R.E. and Miles, D.B. (1994) Ecological and evolutionary inferences from morphology: An ecological perspective. In *Ecological Morphology: Integrative Organismal Biology* (eds. P.C. Wainwright and S.M. Reilly), University of Chicago Press, Chicago, pp. 13–41.

Rohlf, F.J. (1994) *Thin Plate Spline Relative Warps Analysis (TPS), Version 1.22*. Department of Ecology and Evolution, State University of New York at Stony Brook, New York.

Rosenzweig, M.L. (1995) *Species Diversity in Space and Time*, Cambridge University Press, Cambridge, UK.

Roy, K. (1996) The roles of mass extinction and biotic interaction in large-scale replacements: A reexamination using the fossil record of stromboidean gastropods. *Paleobiology*, 22: 436–452.

Roy, K. and Foote, M. (1997) Morphological approaches to measuring biodiversity. *Trends in Ecology and Evolution*, 12: 277–281.

Roy, K., Balch, D.P., and Hellberg, M.E. (2001) Spatial patterns of morphological diversity across the Indo-Pacific: Analysis using strombid gastropods. *Proceedings of the Royal Society of London, Series B*, 268: 2503–2508.

Roy, K., Jablonski, D., and Valentine, J.W. (2004) Beyond species richness: Biogeographic patterns and biodiversity dynamics using other metrics of diversity. In *Frontiers of Biogeography: New Directions in the Geography of Nature* (eds. M.V. Lomolino and L.R. Heaney), Sinauer, Sunderland, MA, pp. 151–170.

Saunders, W.B. and Swan, A.R.H. (1984) Morphology and morphologic diversity of mid-Carboniferous (Namurian) ammonoids in time and space. *Paleobiology*, 10: 195–228.

Sepkoski, J.J., Jr. (1981) A factor analytical description of the Phanerozoic marine fossil record. *Paleobiology*, 7: 36–53.

Sepkoski, J.J., Jr. (1988) Alpha, beta, or gamma: Where does all the diversity go? *Paleobiology*, 14: 221–234.

Sepkoski, J.J., Jr. (1993) Ten years in the library: New data confirm paleontological patterns. *Paleobiology*, 19: 246–257.

Silberling, N.J. and Nichols, K.M. (1982) Middle Triassic molluscan fossils of biostratigraphic significance from the Humboldt Range, northwestern Nevada. *U.S. Geological Survey Professional Paper*, 1207: 1–77.

Skinner, B.J. and Porter, S.C. (1999) *The Dynamic Earth: An Introduction to Physical Geology*, John Wiley and Sons, New York.

Stampfli, G., Borel, G., Cavazza, W., Mosar, J., and Ziegler, P.A. (2001) *The Palaeotectonic Atlas of the Peri-Tethyan Domain*, European Geophysical Union.

Teusch, K.P., Jones, D.S., and Allmon, W.D. (2002) Morphological variation in turritellid gastropods from the Pleistocene to Recent of Chile: Association with upwelling intensity. *Palaios*, 17: 366–377.

Tozer, E.T. (1981a) Triassic Ammonoidea: Classification, evolution, and relationship with Permian and Jurassic forms. In *The Ammonoidea, Systematics Association Special Volume 18* (eds. M.R. House and J.R. Senior), Academic Press, London, pp. 66–100.

Tozer, E.T. (1981b) Triassic ammonoids: Geographic and stratigraphic distribution. In *The Ammonoidea, Systematics Association Special Volume 18* (eds. M.R. House and J.R. Senior), Academic Press, London, pp. 397–431.

Tozer, E.T. (1990) How many Rhaetians? *Albertinia*, 8: 10–14.

Tozer, E.T. (1994) Canadian Triassic ammonoid faunas. *Geological Survey of Canada Bulletin*, 467: 1–562.

Trop, J.M., Ridgway, K.D., Manuszak, J.D., and Layer, P. (2002) Mesozoic sedimentary-basin development on the allocthonous Wrangellia composite terrane, Wrangell Mountain terrane, Alaska: A long-term record of terrane migration and arc construction. *Geological Society of America Bulletin*, 114: 693–717.

Upchurch, P., Slater, C.S.C., and McGowan, A.J. (2006) Meeting background and objectives. *Palaeogeography and Palaeobiogeography: Biodiversity in Space and Time*. NIEeS, Cambridge, UK, http://www.tethys.org.uk/biogeography

Urlichs, M. and Mundlos, R. (1985) Immigration of cephalopods into the Germanic Muschelkalk Basin, and its influence on their suture line. *Sedimentary and Evolutionary Cycles. Lecture Notes in Earth Sciences*, 1: 221–236.

Van Valen, L. (1974) Multivariate statistics in natural history. *Journal of Theoretical Biology*, 45: 235–247.

Villier, L. and Korn, D. (2004) Morphological disparity of ammonoids and the mark of Permian mass extinctions. *Science*, 306: 264–267.

Wagner, G.P. and Altenberg, L. (1996) Complex adaptations and the evolution of evolvability. *Evolution*, 50: 967–976.

Westermann, G.E.G. (1996) Ammonoid life and habit. In *Ammonoid Paleobiology* (eds. N.H. Landman, K. Tanabe, and R.A. Davis), Plenum, New York, pp. 607–707.

Westrop, S.R. and Adrain, J.M. (2001) Sampling at the species level: Impact of spatial biases on diversity gradients. *Geology*, 29: 903–906.

Wills, M.A. (1998) Crustacean disparity through the Phanerozoic: Comparing morphological and stratigraphic data. *Biological Journal of the Linnean Society*, 65: 455–500.

Wills, M.A. (2001) Morphological disparity: A primer. In *Fossils, Phylogeny, and Form: An Analytical Approach* (eds. J.M. Adrain, G.D. Edgecombe, and B.S. Lieberman), Kluwer Academic/Plenum, New York, pp. 55–143.

Wills, M.A., Briggs, D.E.G., and Fortey, R.A. (1994) Disparity as an evolutionary index: A comparison of Cambrian and Recent arthropods. *Paleobiology*, 20: 93–130.

Wilson, K.M., Pollard, D., Hay, W.W., Thompson, S.L., and Wold, C.N. (1994) General circulation model simulations of Triassic climates; Preliminary results. Pangea; Paleoclimate, tectonics, and sedimentation during accretion, zenith, and breakup of a supercontinent. *Geological Society of America Special Paper*, 288: 91–10.

Young, R.E., Vecchione, M., and Donovan, D.T. (1998) The evolution of coleoid cephalopods and their present biodiversity and ecology. *South African Journal of Marine Science*, 20: 393–420.

Index

Systematics Association Publications

1. Bibliography of Key Works for the Identification of the British Fauna and Flora, 3rd edition (1967)[†]
   Edited by G.J. Kerrich, R.D. Meikie, and N. Tebble
2. Function and Taxonomic Importance (1959)[†]
   Edited by A.J. Cain
3. The Species Concept in Palaeontology (1956)[†]
   Edited by P.C. Sylvester-Bradley
4. Taxonomy and Geography (1962)[†]
   Edited by D. Nichols
5. Speciation in the Sea (1963)[†]
   Edited by J.P. Harding and N. Tebble
6. Phenetic and Phylogenetic Classification (1964)[†]
   Edited by V.H. Heywood and J. McNeill
7. Aspects of Tethyan Biogeography (1967)[†]
   Edited by C.G. Adams and D.V. Ager
8. The Soil Ecosystem (1969)[†]
   Edited by H. Sheals
9. Organisms and Continents through Time (1973)[*]
   Edited by N.F. Hughes
10. Cladistics: A Practical Course in Systematics (1992)[‡]
    P.L. Forey, C.J. Humphries, I.J. Kitching, R.W. Scotland, D.J. Siebert, and D.M. Williams
11. Cladistics: The Theory and Practice of Parsimony Analysis (2nd edition) (1998)[‡]
    I.J. Kitching, P.L. Forey, C.J. Humphries, and D.M. Williams

[†] Published by the Systematics Association (out of print)
[*] Published by the Palaeontological Association in conjunction with the Systematics Association
[‡] Published by Oxford University Press for the Systematics Association

SYSTEMATICS ASSOCIATION SPECIAL VOLUMES

1. The New Systematics (1940)[a]
   Edited by J.S. Huxley (reprinted 1971)
2. Chemotaxonomy and Serotaxonomy (1968)[*]
   Edited by J.C. Hawkes
3. Data Processing in Biology and Geology (1971)[*]
   Edited by J.L. Cutbill

4. Scanning Electron Microscopy (1971)*
   Edited by V.H. Heywood
5. Taxonomy and Ecology (1973)*
   Edited by V.H. Heywood
6. The Changing Flora and Fauna of Britain (1974)*
   Edited by D.L. Hawksworth
7. Biological Identification with Computers (1975)*
   Edited by R.J. Pankhurst
8. Lichenology: Progress and Problems (1976)*
   Edited by D.H. Brown, D.L. Hawksworth, and R.H. Bailey
9. Key Works to the Fauna and Flora of the British Isles and Northwestern
   Europe, 4th edition (1978)*
   Edited by G.J. Kerrich, D.L. Hawksworth, and R.W. Sims
10. Modern Approaches to the Taxonomy of Red and Brown Algae (1978)*
    Edited by D.E.G. Irvine and J.H. Price
11. Biology and Systematics of Colonial Organisms (1979)*
    Edited by C. Larwood and B.R. Rosen
12. The Origin of Major Invertebrate Groups (1979)*
    Edited by M.R. House
13. Advances in Bryozoology (1979)*
    Edited by G.P. Larwood and M.B. Abbott
14. Bryophyte Systematics (1979)*
    Edited by G.C.S. Clarke and J.G. Duckett
15. The Terrestrial Environment and the Origin of Land Vertebrates (1980)*
    Edited by A.L. Panchen
16. Chemosystematics: Principles and Practice (1980)*
    Edited by F.A. Bisby, J.G. Vaughan, and C.A. Wright
17. The Shore Environment: Methods and Ecosystems (2 volumes) (1980)*
    Edited by J.H. Price, D.E.C. Irvine, and W.F. Farnham
18. The Ammonoidea (1981)*
    Edited by M.R. House and J.R. Senior
19. Biosystematics of Social Insects (1981)*
    Edited by P.E. House and J.-L. Clement
20. Genome Evolution (1982)*
    Edited by G.A. Dover and R.B. Flavell
21. Problems of Phylogenetic Reconstruction (1982)*
    Edited by K.A. Joysey and A.E. Friday
22. Concepts in Nematode Systematics (1983)*
    Edited by A.R. Stone, H.M. Platt, and L.F. Khalil
23. Evolution, Time and Space: The Emergence of the Biosphere (1983)*
    Edited by R.W. Sims, J.H. Price, and P.E.S. Whalley
24. Protein Polymorphism: Adaptive and Taxonomic Significance (1983)*
    Edited by G.S. Oxford and D. Rollinson
25. Current Concepts in Plant Taxonomy (1983)*
    Edited by V.H. Heywood and D.M. Moore

26. Databases in Systematics (1984)*
   Edited by R. Allkin and F.A. Bisby
27. Systematics of the Green Algae (1984)*
   Edited by D.E.G. Irvine and D.M. John
28. The Origins and Relationships of Lower Invertebrates (1985)‡
   Edited by S. Conway Morris, J.D. George, R. Gibson, and H.M. Platt
29. Infraspecific Classification of Wild and Cultivated Plants (1986)‡
   Edited by B.T. Styles
30. Biomineralization in Lower Plants and Animals (1986)‡
   Edited by B.S.C. Leadbeater and R. Riding
31. Systematic and Taxonomic Approaches in Palaeobotany (1986)‡
   Edited by R.A. Spicer and B.A. Thomas
32. Coevolution and Systematics (1986)‡
   Edited by A.R. Stone and D.L. Hawksworth
33. Key Works to the Fauna and Flora of the British Isles and Northwestern Europe, 5th edition (1988)‡
   Edited by R.W. Sims, P. Freeman, and D.L. Hawksworth
34. Extinction and Survival in the Fossil Record (1988)‡
   Edited by G.P. Larwood
35. The Phylogeny and Classification of the Tetrapods (2 volumes) (1988)‡
   Edited by M.J. Benton
36. Prospects in Systematics (1988)‡
   Edited by J.L. Hawksworth
37. Biosystematics of Haematophagous Insects (1988)‡
   Edited by M.W. Service
38. The Chromophyte Algae: Problems and Perspective (1989)‡
   Edited by J.C. Green, B.S.C. Leadbeater, and W.L. Diver
39. Electrophoretic Studies on Agricultural Pests (1989)‡
   Edited by H.D. Loxdale and J. den Hollander
40. Evolution, Systematics, and Fossil History of the Hamamelidae (2 volumes) (1989)‡
   Edited by P.R. Crane and S. Blackmore
41. Scanning Electron Microscopy in Taxonomy and Functional Morphology (1990)‡
   Edited by D. Claugher
42. Major Evolutionary Radiations (1990)‡
   Edited by P.D. Taylor and G.P. Larwood
43. Tropical Lichens: Their Systematics, Conservation, and Ecology (1991)‡
   Edited by G.J. Galloway
44. Pollen and Spores: Patterns and Diversification (1991)‡
   Edited by S. Blackmore and S.H. Barnes
45. The Biology of Free-Living Heterotrophic Flagellates (1991)‡
   Edited by D.J. Patterson and J. Larsen
46. Plant–Animal Interactions in the Marine Benthos (1992)‡
   Edited by D.M. John, S.J. Hawkins, and J.H. Price

47. The Ammonoidea: Environment, Ecology, and Evolutionary Change (1993)‡
    Edited by M.R. House
48. Designs for a Global Plant Species Information System (1993)‡
    Edited by F.A. Bisby, G.F. Russell, and R.J. Pankhurst
49. Plant Galls: Organisms, Interactions, Populations (1994)‡
    Edited by M.A.J. Williams
50. Systematics and Conservation Evaluation (1994)‡
    Edited by P.L. Forey, C.J. Humphries, and R.I. Vane-Wright
51. The Haptophyte Algae (1994)‡
    Edited by J.C. Green and B.S.C. Leadbeater
52. Models in Phylogeny Reconstruction (1994)‡
    Edited by R. Scotland, D.I. Siebert, and D.M. Williams
53. The Ecology of Agricultural Pests: Biochemical Approaches (1996)**
    Edited by W.O.C. Symondson and J.E. Liddell
54. Species: The Units of Diversity (1997)**
    Edited by M.F. Claridge, H.A. Dawah, and M.R. Wilson
55. Arthropod Relationships (1998)**
    Edited by R.A. Fortey and R.H. Thomas
56. Evolutionary Relationships among Protozoa (1998)**
    Edited by G.H. Coombs, K. Vickerman, M.A. Sleigh, and A. Warren
57. Molecular Systematics and Plant Evolution (1999)‡‡
    Edited by P.M. Hollingsworth, R.M. Bateman, and R.J. Gornall
58. Homology and Systematics (2000)‡‡
    Edited by R. Scotland and R.T. Pennington
59. The Flagellates: Unity, Diversity, and Evolution (2000)‡‡
    Edited by B.S.C. Leadbeater and J.C. Green
60. Interrelationships of the Platyhelminthes (2001)‡‡
    Edited by D.T.J. Littlewood and R.A. Bray
61. Major Events in Early Vertebrate Evolution (2001)‡‡
    Edited by P.E. Ahlberg
62. The Changing Wildlife of Great Britain and Ireland (2001)‡‡
    Edited by D.L. Hawksworth
63. Brachiopods Past and Present (2001)‡‡
    Edited by H. Brunton, L.R.M. Cocks, and S.L. Long
64. Morphology, Shape, and Phylogeny (2002)‡‡
    Edited by N. MacLeod and P.L. Forey
65. Developmental Genetics and Plant Evolution (2002)‡‡
    Edited by Q.C.B. Cronk, R.M. Bateman, and J.A. Hawkins
66. Telling the Evolutionary Time: Molecular Clocks and the Fossil Record (2003)‡‡
    Edited by P.C.J. Donoghue and M.P. Smith
67. Milestones in Systematics (2004)‡‡
    Edited by D.M. Williams and P.L. Forey
68. Organelles, Genomes, and Eukaryote Phylogeny (2004)‡‡
    Edited by R.P. Hirt and D.S. Horner

69. Neotropical Savannas and Seasonally Dry Forests: Plant Diversity, Biogeography, and Conservation (2006)‡‡
   Edited by R.T. Pennington, G.P. Lewis, and J.A. Rattan
70. Biogeography in a Changing World (2006)‡‡
   Edited by M.C. Ebach and R.S. Tangney
71. Pleurocarpous Mosses: Systematics & Evolution (2006)‡‡
   Edited by A.E. Newton and R.S. Tangney
72. Reconstructing the Tree of Life: Taxonomy and Systematics of Species Rich Taxa (2006)‡‡
   Edited by T.R. Hodkinson and J.A.N. Parnell
73. Biodiversity Databases: Techniques, Politics, and Applications (2007)‡‡
   Edited by G.B. Curry and C.J. Humphries
74. Automated Taxon Identification in Systematics: Theory, Approaches, and Applications (2007)‡‡
   Edited by N. MacLeod
75. Unravelling the algae: the past, present, and future of algal systematics (2008)‡‡
   Edited by J. Brodie and J. Lewis
76. The New Taxonomy (2008)‡‡
   Edited by Q.D. Wheeler
77. Palaeogeography and Palaeobiogeography: Biodiversity in Space and Time (in press)‡‡
   Edited by P. Upchurch, A. McGowan, and C. Slater
78. Climate Change, Ecology, and Systematics (2011)§
   Edited by T.R. Hodkinson, M.B. Jones, S. Waldren, and J.A.N. Parnell
79. Biogeography of Microscopic Organisms: Is Everything Small Everywhere? (2011)§
   Edited by D. Fontaneto

  * Published by Clarendon Press for the Systematics Association
  * Published by Academic Press for the Systematics Association
  ‡ Published by Oxford University Press for the Systematics Association
  ** Published by Chapman & Hall for the Systematics Association
  ‡‡ Published by CRC Press for the Systematics Association
  § Published by Cambridge University Press for the Systematics Association

Printed and bound by CPI Group (UK) Ltd, Croydon, CR0 4YY

21/10/2024

01777107-0003